T0205369

PERCEIVING ANIMALS

Perceiving Animals

Humans and Beasts in
Early Modern English Culture

Erica Fudge

Lecturer in English Literary Studies
Faculty of Humanities and Cultural Studies
Middlesex University

First published in Great Britain 2000 by
MACMILLAN PRESS LTD
Houndmills, Basingstoke, Hampshire RG21 6XS and London
Companies and representatives throughout the world

A catalogue record for this book is available from the British Library.

ISBN 978-0-333-72812-3

First published in the United States of America 2000 by
ST. MARTIN'S PRESS, INC.,
Scholarly and Reference Division,
175 Fifth Avenue, New York, N.Y. 10010

ISBN 978-1-349-62417-1   ISBN 978-1-349-62415-7 (eBook)
DOI 10.1007/978-1-349-62415-7

Library of Congress Cataloging-in-Publication Data
Fudge, Erica.
Perceiving animals : humans and beasts in early modern English
culture / Erica Fudge.
p. cm.
Includes bibliographical references and index.

1. Philosophical anthropology—History.   2. Human beings—Animal
nature—History.   3. Animals (Philosophy)—History.   I. Title.
BD450.F7945 1999
179'.3—dc21   99–15589
                                                            CIP

This book is printed on paper suitable for recycling and made from fully managed and
sustained forest sources.

10   9   8   7   6   5   4   3   2   1
09 08 07 06 05 04 03 02 01 00

To my parents, Heather and Erik Fudge

Contents

Illustrations

Acknowledgements

During the researching and writing of this book I have been very lucky to have received help and support from many colleagues, friends and family members who have all been generous with time and references, and who have shown immense patience. The project began under the stimulating and thoughtful supervision of Michael Hawkins at the School of English and American Studies, University of Sussex. I am grateful for his generosity. A complete draft of the manuscript was read by Sue Wiseman. Her comments, ideas, references and good humour have pulled out some of the horrors which lurked at its heart. I am very lucky to have such an insightful friend.

Various chapters were read by kind friends and colleagues. I am grateful for all of the time and effort they put in. Thanks to Nicola Bown, a friend who always listened and came up with good ideas from the beginning to the very end of the project; Ruth Gilbert who was generous with ideas; James Knowles, who has taught, advised and helped me with great kindness over a frighteningly long period of time; Gareth Roberts who made some useful suggestions and sent me off to the library with some new references, and whose advice and friendship I will miss; and Nigel Smith and Wendy Wheeler, who both gave up their time to read chapters. I hope that I have been able to take on their suggestions. Alan Sinfield and Kate McLuskie were generous examiners and have remained thoughtful presences over the years. Tracey Hill and Alan Marshall read parts of the book and offered helpful suggestions. I am also grateful to The Faculty of Humanities, Bath Spa University College for allowing me the time to complete the book.

As well as these friends and colleagues I wish also to thank for various reasons Mark Adams, Amanda Boulter, Carolyn Burdett, Nicola Chambers, Gill Clayborough, Karen Gale, Lisa Gamsby, Jennie McCabe, Alison McInnes, Paul McSorley, Chris Mounsey, Kathryn Perry, and all the contributors to *At The Borders of the Human*.

I would like to thank the librarians and staff at the University of Sussex Library, the British Library, Senate House Library, the Public Record Office (Chancery Lane) and the East Sussex Record Office (Lewes) for their help during the writing of this book. Charmian

Hearne at Macmillan has been a patient and thoughtful editor who had faith in the project and has guided it through from a very early stage. I am indebted to her for her support. Christina Zaba cast an expert eye over the final version, for which I am very grateful. As ever, any remaining mistakes are mine.

Finally thanks go to my family. To Tim, Julie and Tessa for not laughing too much. Most of all I am grateful to my parents, Heather and Erik Fudge: first for the name, and second for their support which came in many forms. This book is dedicated to their kindness.

I am grateful to the publishers for the right to reproduce the following:

Part of Chapter 4 appeared in 'Calling Creatures By Their True Names: Francis Bacon, The New Science and the Beast in Man', in Erica Fudge, Ruth Gilbert, Susan Wiseman (ed.), *At the Borders of the Human: Beasts, Bodies and Natural Philosophy in the Early Modern Period* (Basingstoke: Macmillan, 1999).

Part of Chapters 3 and 4 appeared in 'Temples of God: William Prynne and the New Science', in Tracey Hill and Jeff Rodman ed., *The Body of Truth: Corporeality and Power in Early Modern Culture* (Bath: Sulis Press, 1999).

Part of Chapter 2 appeared in 'Pocahontas' Baptism: Reformed Theology and the Paradox of Desire', *Critical Survey*, 11/1 (1999).

Introduction: The Dangers of Anthropocentrism

There was a Bear Garden in early modern London. In it the spectators watched a pack of mastiffs attack an ape on horseback and assault bears whose teeth and claws had been removed. People enjoyed the entertainment. We know this from the numerous reports of the baitings which have survived. What we don't understand is the nature of their enjoyment. This book began as an attempt to comprehend the pleasure through an examination of the ways in which the spectators related to animals, those silent and, until recently, forgotten creatures of history.[1] What emerged from my reading surprised me. An anxiety could be traced in the ways in which animals were represented: an anxiety which was not about the animals. My attempt to read the Bear Garden revealed a struggle more significant than the one played out by the dogs and the bears, it revealed a struggle over the nature of being human itself.

Achieving human status has never been easy. The ways in which being human is defined rely on the exercise of certain skills, such as speaking, and in the exercise of human-ness the animal becomes an important player in history. It becomes the thing which the human is constantly setting itself against. Because of this *Perceiving Animals* is a book not so much about animals as about the ways in which humans define themselves as human in the face of the animal. The making of the boundary which separates the human from the beast is important in two ways: because it is an issue in many areas of culture which are central to our understanding of the early modern period and because it raises ethical and political issues which remain relevant today. But to read for the human through the representation of the animal in early modern literature is also to open up theoretical problems.

Stephen Greenblatt's metaphorical representation of the new historicist project can still be used to sum up an important element of the historical endeavour: 'I began with the desire to speak with the dead.' The loss of objectivity, the actual longing, which is proposed is an emotive and seductive model of the task of the historian. Greenblatt speaks of the 'textual traces' left by the dead which he

1

wishes to retrieve in order to 're-create a conversation' with the past.[2] But this romantic ideal of the communion of voices from the past and the present is shattered by the end of Greenblatt's theorising:

> I had dreamed of speaking with the dead, and even now I do not abandon this dream. But the mistake was to imagine that I would hear a single voice, the voice of the other. If I wanted to hear the voice of the other, I had to hear my own voice. The speech of the dead, like my own speech, is not private property.[3]

A voice, Greenblatt argues, never speaks alone, a listener never hears in isolation.

But when I began with the desire to speak with the dead at the onset of this book (and, despite my misgivings about new historicism, I know that I did) the project failed in one significant way: the dead I wanted to speak to – the animals – were, unlike Greenblatt's dead, private property, but still they had no voices and left no textual traces. Or rather, the only traces of the animal within the written materials which I had access to were the vellum on which the words were written and the leather in which the books were bound.[4] This absent-presence of animals was firmly, and ironically, underlined for me on the label recording the British Library's 'Record of Treatment, Extraction, Repair, etc.' stuck into the back cover of Thomas Taylor's *A Vindication of the Rights of Brutes* (1792). In the column next to the word 'Adhesives' is written by hand 'ANIMAL GLUE'.[5]

This treatment dates from 1996, but the treatment of animals in written material has always been similarly bound up with animals as absent-presences: there, but not speaking. In this sense speaking with the dead became impossible on two scores: animals do not speak my language, and they do not write, leave textual traces, other than the traces – vellum, leather, glue – which speak of their objectification. I had to look to humans to find the animals, but all that is available to the historian are records of use, edibility, training, exploitation. When John Simons says that the project of 'Cultural Studies after Speciesism' must be 'reoriented towards ... the monkeys not the organ grinders'; he recognises the necessity of skewing the old vision, but the problem is, of course, that we have no access to monkeys without going first of all through organ grinders.[6] In historical terms the animal can never be studied in isolation, it is always a record by and of the human. This is the reason why this book undermines the apparently antithetical binary of animal and human. It is about both animals and humans; the link is inevitable. Reading about animals is always reading through humans, and, as later chapters will show, reading about humans is reading through animals.

This sense of animals as absent and yet present can be traced in a brief look at three poems from the early modern period. Ben Jonson's representation of the natural world in 'To Penshurst' offers an image of unmitigated and unquestioned human dominion. In his panegyric the benificence of the householder, Sir Robert Sidney, Viscount Lisle, is paralleled in his power over the natural world: the pheasant is 'willing to be kill'd', the carps 'runne into thy net', the eels 'leape on land, / Before the fisher'.[7] The Edenic perfection of Sidney's world – and with that his absolute mastery – is central. In Thomas Carew's later country-house poem 'To Saxham' a similar dominion emerges. Once again, nature is represented sacrificing itself to humanity:

> The pheasant, partridge, and the lark
> Flew to thy house, as to the Ark.
> The willing ox, of himself came
> Home to the slaughter, with the lamb,
> And every beast did thither bring
> Himself to be an offering.[8]

Human mastery is figured through the animal. In a different context George Herbert offers a comparable image of nature. In 'Providence' he writes:

> Of all the creatures both in sea and land
> Onely to Man thou hast made known thy wayes,
> And put penne alone in his hand,
> And made him Secretarie of thy praise.
>
> Beasts fain would sing; birds dittie to their notes;
> Trees would be tuning on their native lute
> To thy renown: but all their hands and throats
> Are brought to Man, while they are lame and mute.

For Herbert, the animals are incapable of praising God; this is how they are distinguished from man and is how man's centrality in the universe is emphasised. There is a duty to the animal – humans must praise God on their behalf – but the duty to animals places humans at the centre of the world. To underline this anthro-pocentrism even further, while Herbert's animals cannot speak to the divine they can speak to the god on earth, man: 'The beasts say, Eat me.'[9] Their servitude underlines human centrality.

To say that any of these poems is about animals is to miss the point. But animals are present in a way which should not be ignored. Even in these extreme and poetic visions of dominion can be traced an understanding of the relation to the natural world. Animals represent human power: their self-sacrifice is an image of man's control. The poems are not about animals but are, in part, about the human sense of what an animal should be. For this reason the ideas voiced in 'To Penshurst', 'To Saxham' and 'Providence' are a part of this project.

But the animals in these poems need also to be understood in another and more problematic way. In running into nets, willingly offering themselves for sacrifice, asking for human aid, the animals are fulfilling human ideals, but are fulfilling them in anthropomor-phic ways. The ox knows that it is to be eaten, it has an understand-ing of its duty, in the same way that the poet has an understanding of his duty when he writes the poem to his patron, or to God. Anthropocentrism creates anthropomorphism: for the ox to be willing it must have a will. Where human power over animals is represented it often undercuts humanity as a separate category.

More generally, in writings dealing with the animal in early modern England – whether theological, humanist, scientific or legal – the animal is represented as the antithesis of the human. But in presenting the animal as the thing which the human is not – begging to be eaten, for example – writers give animals a status, that of beggar, which undermines the desire to make a clear separation between the species. To assert human supremacy writers turn to discuss animals, but in this turning they reveal the frailty of the supremacy which is being asserted. Paradoxically, humans need animals in order to be human. The human cannot be separated because in separation lies unprovability.

In *The Gay Science* Friedrich Nietzsche warns 'Let us beware of thinking that the world is a living being.' Aphorism 109 spells out the dangers of anthropomorphism; attributing to the world the qualities of 'order, arrangement, form, beauty, wisdom'.[10] For Nietzsche anthropomorphism is an absurdity; he asks, 'Where should [the world] expand? On what should it feed? How should it grow and multiply?'[11] Our linguistic representations limit us to an understanding which creates the world in our own image which in turn presents the world as having very human frailties. But we do not need to look only to the late nineteenth century for this inter-pretation. A similar sense of the restricted nature of human vision can be traced in the work of Francis Bacon, as he writes in *Novum Organum*:

> it is a false assertion that the sense of man is the measure of things. On the contrary, all perceptions, as well as the sense as of the mind are according to the measure of the individual and not according to the measure of the universe. And the human under-standing is like a false mirror, which, receiving rays irregularly, distorts and discolours the nature of things by mingling its own nature with it.[12]

The '*Idols of the Tribe*', as these anthropomorphisms are called, are natural to the human mind and therefore impossible to dismiss. Humans will always place themselves at the centre of 'their' world. Anthropocentrism is inevitable.

In his preface to *Philomythie or Philomythologie*, entitled 'Sarcasmos Mvndo', Thomas Scot pre-empts both Nietzsche and Bacon. Not only does he present in a couplet a version of the anti-teleological Aphorism 112 from *The Gay Science* – 'Causes foregoe effects by cause of kinde, / Yet first th'effect and then the cause we find.' He also mocks the very anthropomorphisms which both Bacon and Nietzsche regard as an inevitable part of human thought:

> Attend my Doctrine then. I say this Earth
> On which we tread, fro[m] whence wee take our birth,
> Is not, as some haue thought, proportio[n]'d rownde,
> And *Globe-like* with such zones, & girdles bounde.
> As *Poets* or (more lyers) *Trauellers* say,
> But shaped awry, and lookes an other way.
> It is a monstrous Creature like a Man,
> Thrust altogether on a heape, wee can
> Distinguish no part, goggle eyes, wide mouth
> Eares that reach both the poles fro[m] north to south,
> Crump shouldered, breast, & backe, & thighes together,
> The leges and feet all one, if it hath either.

.... And when he stirres a limme, or breaketh wind,
We cal't an earth-quake, and the danger find.
Kings, Emperours, and mighty men that tread
In highest state, are lyce vpon his head.
The *Pope* and all his traine are skipping fleas
That know no bounds, but leap lands-law & seas.
The rest are nits or body lice, that craule
Out of his sweate, such vermine are we all.[13]

The verse illustrates the logical outcome of anthropocentrism, but it also mocks the failings of this world view. In seeing the world as shaped like a man we can 'distinguish no part', all we can do is fabricate meanings: the earthquake is a name given to the unknowable disruption of the natural order, is the human 'excuse' for what is perceived as a global fart. But mockery is not all that the verse proposes. By taking anthropomorphism to its extreme Scot recognises the result. If the world is like man, then man must be like a flea. By seeing the world like ourselves we reduce ourselves to the thing we desire not to be. This is not merely a literary trope: the issues in Jonson, Carew, Herbert and Scot's work also raise ethical questions.

A sea anemone puts out its tentacles to trap the passing microorganisms, its food. The tentacles, as anyone who has ever poked around in a rockpool will know, are retracted as a defence mechanism. In describing the actions of the sea anemone we say that it is hungry, or that it defends itself. Nietzsche, Bacon and Scot all ask, in very different ways, if these terms are appropriate for an organism as simple as this. Do we really want to attribute the notion of 'hunger' to an anemone? Is this basic organism really feeling something akin to what we feel? The questions might sound ridiculous, but the answers they elicit raise important and complex problems. If we say no, the anemone obviously does not feel hunger as we understand the term, then we lose the ability to describe this creature. We have no other ways to explain the movement of the tentacles and the anemone is excluded from our thought environment. But if we accept that the term 'hunger' is applicable then we are giving the creature a capacity which is like our own. We are including it in our world, and new problems arise. Put simply, if a beast can say 'Eat me' it is no longer edible.

This undercutting of the distinction between the human and the animal comes to the fore in a different way in Diana Fuss' discussion of Nietzsche's representation of thinking as rumination in

On the Genealogy of Morals – 'one has', he says, 'almost to be a cow'. Fuss wonders, 'Is Nietzsche asking us to ponder more than simply the anthropomorphic notion that cows may think but also the far more disarming possibility that thinking may itself be a bovine activity?'[14] If thinking, reasoning, the thing which is so often used to distinguish the human from the animal, is bovine, where is the distinction? Where is the human? Nietzsche's response to this was to propose the figure of the Ubermensch, the superman who would rise above the human, but who would also be no longer able to communicate with humans, and as communication is another proof of humanity this defeats the notion of the human. The alternative in Nietzsche's thought is the indistinguished, and indistinguishable masses. The human (all too human) becomes a member of the herd; becomes an animal. Humanity, as a separate and distinct category, becomes impossible.

But anthropocentrism and anthropomorphism are not merely issues within a theoretical space. They have practical applications. The destruction of humanity is not merely a deconstructive concept; it has real and painful results, something which David Clark makes clear in his reading of Emmanuel Levinas' 'The Name of a Dog, or Natural Rights'. Writing of his experience in Camp 1492 Levinas recalls that he and his fellow Jews were 'subhuman, a gang of apes' in the eyes of the Nazi guards. Into this appalling environment comes a dog, Bobby: 'He would appear at morning assembly and was waiting for us as we returned, jumping up and down and barking in delight. For him, there was no doubt that we were men.'[15] Bobby's ability to recognise the humanity of the Jewish prisoners places him above the guards, but Levinas refuses, cannot bear, to anthropomorphise the dog – 'enough of allegories!' he writes – because, as Clark notes, 'the sentimental humanization of animals and the brutal animalization of humans are two sides of the same assimilating gesture. In humanizing the animal, these fictions risk the tropological reversal by which persons are in turn bestialized'.[16] If he accepts that Bobby is capable of something akin to human recognition then Levinas must also accept the dehumanisation of himself and the other prisoners. Bobby's presence is a comfort, but it is also a threat.

With this in mind, if anthropocentrism – placing the human and human vision at the centre – leads, as I have argued, to anthropomorphism – seeing the world in our own image – and anthropomorphism allows for the animalisation of humans then anthropocentrism paradoxically destroys *anthropos* as a category. By centralising the human, making the human vision the only vision, the separation of the species is impossible. At the heart of the debate about animals lies a debate about humanity which has social and political ramifications. If an animal can beg, then is a (human) beggar also an animal? The implications of this question are played out in the sense that in order to assert human status writers have to make exclusions. Some humans are aligned with animals: in fact, some humans are not human at all.

In *Man and the Natural World* Keith Thomas traced the changes in the ways in which humans understood the natural world over a period of 300 years. Rather than taking Thomas' broadly chronological approach – starting with the theological foundations of early modern ideas, and ending with the increasing debates about vegetarianism and the ethics of human dominion in the seventeenth and eighteenth centuries – I look primarily at the ways in which different discourses represent the animal. The first chapter begins in the Bear Garden and offers a microcosm of the debates which the following chapters develop. The depth of the destruction of human as a species which stands alone, which can assert without problem its superiority, becomes clear through an examination of baiting. The rest of the book takes up this idea and inserts the animal into what are canonical debates about Reformed ideas in the church,[17] theories of education, the emergence of the new science, the systematisation of the common law, and Leveller political ideas. What emerges upsets our concept – often unthought, unproblematised – of the human.

An underlying organising principle of this book is the development of the human itself: from the infant and baptism in the second chapter, through, in succeeding chapters, education, linguistic control and the inheritance of property to, finally, enfranchisement. The stages of life – I offer five, rather than Jaques' seven – reveal an image of a changing notion of the human and of what constitutes being human. The organisation of the book around these different stages of life moves chronologically through the period 1558–1649. After beginning with a visit to the Bear Garden in 1562 in the first chapter, the second concentrates on the establishment of Reformed religious principles which was taking place during the reign of Elizabeth I. Succeeding chapters deal with issues which coincide with and postdate these. Chapter 3 looks at the work of Sir Philip Sidney and Ben Jonson and concentrates on the period 1570–1610.

Chapter 4 traces the impact of Francis Bacon, whose final work, *The New Atlantis*, was written in 1624 and first published in 1627. The work of Edward Coke, which spans a similar period to Bacon's, is central to Chapter 5, and finally, in Chapter 6, I look at the writings of Richard Overton which were published during the period 1643–9. At the end we have arrived at the establishment of the Commonwealth. This concluding date is not merely fortunate in that it closes the book just as massive changes were taking place, it is also appropriate because what has emerged in Overton's writings is, I will argue, a revision of what has gone before in the ideas of Perkins, Sidney and Jonson, Bacon and Coke. It is in the social and political realm, as Levinas' experiences reveal so acutely, that the destruction of humanity finds its true end. But it is also through politics that an attempt to reinstate the category can take place. In his call for revolution Overton is not only asking for social and political change, he is asking for natural change. Finally, in the Epilogue, I return to the Bear Garden and to *Overton's Defyance of the Act of Pardon*. What is revealed in this text is defeat and degradation: the human is still animal and the baiting persists.

The terms used in this book need some form of explanation. I continue to use the term human, even as I argue that in early modern England it is becoming unstable, partly because there is need of a term to represent the thing which is destroyed, and partly because it is a term which is constantly in use in the texts which I analyse. Even while they reveal the difficulty of being human the notion never disappears. Similarly, I often use the term 'man', not as an unthought shorthand for human, but as a deliberate reminder of the gendering of even the most general invocations of the species in this period.

An alternative to the term 'human' which avoids some of its problems is employed here: that is, 'human-ness'. This is used to represent the qualities which, I argue, each area of thought proposes as specific to the human. In William Perkins' Reformed ideas, for example, it is the operation of conscience which represents human-ness. Without conscience, Perkins implies, there is no human. Human-ness is a useful category as it reveals the divisions which exist between being a human and possessing human qualities.

'Humanity' is an even more problematic term than 'human', for within the materials read here one of the recurring themes is of exclusion. These exclusions are based on qualities of human-ness;

the having or not having of, for example, a conscience. Humanity, in this context, represents the unproblematic and unthought category of human: those who are human whether or not they possess the qualities of human-ness. It is a term which is used to remind of the exclusive nature of human-ness in early modern writing. All of humanity does not necessarily possess human-ness, and the possession of human-ness does not necessarily imply human status. Being human is not a given, it is achieved. Margaret T. Hodgen has noted that the term 'race' in the early modern period 'carried a zoological connotation properly applicable only to animals'. The displacement of the term to designate different peoples was a development of early modern thought. 'New World man or the naked and threatening savage took that place in thought which, during the Middle Ages, had been reserved for human monsters. If human their's was a degraded humanity.'[18] The degradation of humanity in the face of the beast in early modern thought is a recurring theme.

Throughout the book I also use the term 'species'. According to the *OED*, species gained the meaning which we now understand during the seventeenth century. In the late sixteenth century species meant 'A distinct class, sort, or kind, of something specifically mentioned or indicated.' This did not relate only to the organisation of the natural world, but could relate to any group of ideas or objects. The more recognisable use of the term, representing a 'group or class of animals or plants (usually constituting a subdivision of a genus)' is a later addition to the language, but is one which I use here to distinguish more generally the human from the animal.[19] This is an application of an anachronistic notion but reflects, I suggest, the ways in which writers were distinguishing between themselves and beasts.

But the starting-point is in the Bear Garden, the most explicit and spectacular arena of the abuse of animals. I begin with a reading of baiting which asserts the status of human as unproblematic – as always-already – and then show how such a reading fails to respond to the complications which exist in early modern culture. A variety of texts from the areas with which I am later concerned are used to propose the thoroughgoing nature of the loss of human status in early modern England. At the end of the chapter two emblems are read which offer a brief glimpse of the issues at stake in this period. The aim of the first chapter is to offer in miniature some of the problems with which the rest of the book deals and to reveal the depth to which the category human was under threat between 1558 and 1649.

1

Screaming Monkeys: The Creatures in the Bear Garden

I

In 1562 Alessandro Magno, an Italian merchant's son, went to the Bear Garden in London. It was one of the many tourist attractions he visited during his stay and his commentary on the entertainment he witnessed on the Bankside remains one of the most detailed we have:

> Let me explain that first they take into the ring – which is fenced around, so that one cannot get out unless the gate is opened – a cheap horse with all his harness and trappings, and a monkey in the saddle. Then they attack the horse with five or six of the youngest dogs. Then they change the dogs for more experienced ones. In this sport it is wonderful to see the horse galloping along, kicking up the ground and champing at the bit, with the monkey holding very tightly to the saddle, and crying out frequently when he is bitten by the dogs. After they have entertained the audience for a while with this sport, which often results in the death of the horse, they lead him out and bring in bears – sometimes one at a time and sometimes altogether. But this sport is not very pleasant to watch.[1]

The distinction made by Magno between the monkey-baiting, which he enjoys, and the bear-baiting, which is 'not very pleasant', seems incomprehensible. But the qualitative difference which Magno sees can tell us much about one way in which baiting was understood.

Magno's image of the monkey 'crying out' (in another translation of this passage the monkey 'screams'[2]) is obviously and disturbingly

11

anthropomorphic. There is a sense of recognition: the monkey is a creature similar to the human. Even to the pre-Darwinian sensibilities of the early modern period the link between human and ape is very clear.[3] In one (very long) sentence Edward Topsell illustrates the dangerously ambivalent nature of the creature:

> Apes do outwardly resemble men very much, and *Vesalius* sheweth, that their proportion differeth from mans in moe things then *Galen* obserued, as in the muscles of the breast, & those that moue the armes, the elbow and the Ham, likewise in the inward frame of the hande, in the muscles mouing the toes of the feete & the feete and shoulders, & the instrument mouing in the sole of the foote, also in the fundament & mesentary, the lap of the liuer, & the hollow vain holding it vp, which me[n] haue not; yet in their face nostrils, eares, eye-lids, breasts, armes, thumbes, fingers & nailes, they agree very much.[4]

The ape is both like and not like the human, but ultimately it is anthropoid; its face is like a human face.

In the Bear Garden the animal was on horseback and may well have been dressed to underline its likeness to the human.[5] By staging the anthropoid nature of the animal in such an obvious way the spectator was invited to perform two forms of recognition: to recognise the anthropoid nature of the animal, but also to recognise that anthropoid only ever means human-like, it can never mean human. At the moment of sameness difference is revealed and the disturbing spectacle of the screaming monkey on horseback becomes a reminder of the superiority of humanity. The monkey can only ever achieve a comic imitation of the human.[6]

There are parallels between Homi Bhabha's work on colonial literature and this reading of monkey-baiting. Bhabha discusses the notion of 'Anglicizing', noting that the Indian native is only ever Anglicised, never made English, and that 'to be Anglicized, is *emphatically* not to be English.'[7] Similarly, in the Bear Garden, to be anthropoid is *emphatically* not to be human. It might seem somewhat crude to make such an analogy between the animal and the native other but, historically at least, attitudes to the two groups have been frighteningly similar as Marjorie Spiegel has noted in her aptly titled book *The Dreaded Comparison*.[8]

Peter Stallybrass and Allon White have argued that animals in fairground booths in the early modern period consolidate 'the sense that the civilized is always-already given'.[9] The parody of human behaviour which is traced in the animals' comic attempts to perform human actions can only work if there is a clear and secure sense of what is correct human behaviour. Stallybrass and White's account is clearly replicated in this interpretation of the Bear Garden: the monkey on horseback reinforces the status of the human viewer. In *Bull, Beare, and Horse*, a comic verse dedicated to Thomas Godfrey 'Keeper of the Game for Beares, Bulls, and Dogges', John Taylor the Water Poet writes with irony of how the audience watching 'Jack an Ape' might 'see and learne some courage from a Beast'. The mock-heroic nature of the poem parodies contemporary defences of cock-fighting,[10] and underlines the comic nature of the ape on horseback:

> Where *Jack-an-Apes* his horse doth swiftly run
> His circuit, like the horses of the sun,
> And quicke as lightning, hee will trace and track,
> Making that endlesse round his Zodiake,
> Which *Jacke* (his Rider) bravely rides a straddle,
> And in his hot Careere perfumes the saddle.[11]

The ape might *resemble* a figure from mythology, but ultimately resemblance – like the notion anthropoid – is shown to be very different from the thing itself: he pisses in the saddle. This perhaps explains the pleasure which Magno experiences watching the monkey-baiting: his sense of his own humanity is constantly being reinforced. The monkey is both like him and not like him, and it is in the comedy of such a spectacle that entertainment is found as the human remains a stable category throughout. In fact, the stability of the category would seem to be the main result of the sport.

The bears, however, present Magno with a different spectacle because they represent an alternative notion of the natural world. This part of the entertainment is a representation of wild nature controlled and choreographed (in the main) by humans. But with this sense of human dominion there is also an alternative sense of danger. The spectators 'cannot get out unless the gate is opened', for their safety they are trapped in the Bear Garden just as the bear is chained to the stake, and this serves as a reminder that, although it is the bears who are apparently confined by humans, this confinement is replicated in the fencing in of the spectators.[12] Human dominion, the Bear Garden seems to say, has its limits; and this is perhaps the most potent attribute of bear-baiting. There is power, but power brings with it danger. Baiting has a moral function. The spectacle which proposes so explicitly in one instance the strength of humanity's claim to superiority through the difference between anthropoid and *anthropos* in another reminds humans of their weakness in the face of wild nature, and for this reason the bears' fight with the dogs is, for Magno, 'not very pleasant to watch'.

This sense of baiting as both didactic and dangerous is repeated in some religious works from the early modern period. Phillip Stubbes regarded most amusements of the late sixteenth century as potential threats to the human soul, and baiting was a clearly defined form of blasphemy:

> What Christia[n] harte can take pleasure to see one poore beast to rent, teare, and kill an other, and all for his foolish pleasure? And although thei be bloudie beasts to mankind, and seeke his destruction, yet we are not to abuse them, for his sake who made them, and whose creatures thei are. For notwithstandyng that thei be euill to vs, & thirst after our bloud, yet are thei good creatures in their own nature and kind, and made to set forth the glorie, power and magnificence of our God, and for our vse, and therefore for his sake wee ought not to abuse them.[13]

The animals have a function – setting forth God's glory. This is a function which is mirrored in Herbert's later declaration of anthropocentrism in 'Providence' and it is also to be traced in other religious ideas. William Perkins, like Stubbes, saw bear-baiting as 'no meet recreation'. He argued that 'the antipathie and crueltie, which one beast sheweth to another, is the fruite of our rebellion against God, and should rather mooue vs to mourne, then to reioyce'.[14] For Perkins the cruelty exhibited in these sports is not human but is manifested by the animals involved in the combat. However, the central issue is not the suffering of the animals but the reminder which the sport gives of the essential depravity of postlapsarian humanity. The ability to abuse animals is evidence of the Fall. Cruelty is a reminder of the status of man. But the didactic function of baiting emerges with more dangerous implications in Robert Bolton's work thirty years later:

> Bathe not thy recreations in blood: Refresh not thy tired mind with spectacles of crueltie: Consider, 1. How God himselfe out of

tendernesse and pittie, would not haue his people feede vpon the flesh of Beasts with the blood, lest thereby they should be flesht to crueltie, and inured to behold rufull obiects without horrour. And doest thou thinke then, hee will allow thee to feede thine eye and fancy with their bloody torturing and tearing one another in pieces? 2. With what brutish sauagenesse thou dejectests and debasest humanitie, below the immanitie of beasts. No beast, they say, takes contentment in the hurting of any other, except in the case of hunger or anger. They satisfie their appetites and rage sometimes with cruelty and blood; but their eyes and fancies neuer. 3, That men bloodily minded towards harmlesse beasts, discouer our naturall propension to crueltie ...[15]

To watch a cruel entertainment such as baiting is to reveal the truth about humans. They sink below the level of the beasts. This is where this reading of the Bear Garden as a place where being human is reinforced becomes problematic. The reading of what baiting might achieve which utilises Stallybrass and White's notion of the always-already human is very different from what is perceived by contemporary commentators and spectators. There is no always-already human as a synonym for civilised (with civilisation a synonym for human-ness), there is instead a human with 'a naturall propension to crueltie'. To watch a baiting, to enact anthropocentrism, is to reveal, not the stability of species status, but the animal that lurks beneath the surface. In proving their humanity humans achieve the opposite. The Bear Garden makes humans into animals. This different interpretation of the meaning of baiting raises new problems for the status of humanity and another way of understanding the entertainment is needed.

II

In her discussion of the early twentieth-century anti-vivisection movement Coral Lansbury touches on the meaning of baiting and her reading is interesting here.

Rather than being seen as an aberration of human nature, the torture and killing of animals permitted those who had no rights, no possibility of ever imposing their will upon others, to demonstrate, often publicly, their strength and dominance. When men who were accustomed to being thrashed and abused could watch

the chained bull harried by a pack of dogs, it was like seeing the
authority of the master torn apart by the mob.[16]

Placing the bull in the role of the authority figure is an interesting
suggestion which is supported by theories of the carnivalesque
which argue for the temporary inversion of social norms.[17] Bull-
baiting is seen as an emblem of revolution, with the dogs representing the lower orders of society. However, this reading does not fit
so well with the sport of monkey-baiting with its obviously anthropoid 'victim', nor with bear-baiting in early modern London where
most of the bears had 'human' names: Harry Hunkes, George Stone
etc.[18] E. P. Thompson's description of eighteenth-century power
and class relations offers an alternative view. He states that 'once a
social system has become "set" it does not need to be endorsed daily
by exhibitions of power (although occasional punctuation of force
will be made to define the limits of the system's tolerance)'.[19] It is
only what is fragile within the social and political system that needs
to be staged, what is in place need not be advertised. Following this,
the show of aggression in a baiting would become a recognition of
the dangers which existed within the social structure. The display
was a reminder to both monarch and plebian that harmonious social
relations could not be taken for granted but had to be worked at.

But in such interpretations, as in religious writings, the animal
itself is lost, or is at best secondary. The cruelty of baiting becomes
symbolic of oppressive power relations. In addition, the possible
emblematic meaning of baiting as a carnivalesque display of political upheaval reveals that the symbolic narratives of the bull/bear as
monarch or the bull/bear as plebian do not work, or rather, that
they can only work as very partial, selective readings, each potentially meaning its exact opposite.[20] Baiting is both undermining and
supporting the social order, it is a performance meaning different
things to different viewers. There is, once again, a refusal to recognise the entertainment as meaningful in and of itself: the animal
disappears from view and what remains is politics, humans. I want
to understand the meaning of the animal in baiting.

Marjorie Spiegel's account of cruelty to animals offers a possibility
of understanding the pleasures of baiting, returning as it does to the
materiality of cruelty. She acknowledges that the place of the
animal is synonymous with the place of some humans:

> Often, when someone is ill-treated or relegated to a demeaning
> position in society, they will respond by venting their frustration

on someone whose societal position is even lower than their own. It is not rational: their violent action in no way serves as a retaliation towards their oppressors. ...

Taking this concept one step further, we can see that by torturing or dominating a powerful animal ... the oppressors feel, unconsciously, that they have destroyed those who hold power over them.

Spiegel then extends this reading to the issue of cruelty:

By destroying or tormenting the weak, such as a rabbit or a child, the oppressors become the master who has in turn tortured them. Their own victims' helpless writhings echo what they have felt, and temporarily replace them in the role of victim. And so these new reactive torturers ascend, momentarily in their own mind, to the social – or physical – power position of their oppressor.[21]

Cruelty to animals is understood as the infliction of pain on those even lower than you as a response to the frustration caused by social inequalities. Baiting represents a response to social problems. This might sound like a continuation of the symbolic readings of baiting which remove the animal from the frame, but in this account the sport possesses a meaning which recognises the inseparability of the natural and political order.

But Spiegel's interpretation of cruelty can also be turned back on itself and the implications of the anthropocentrism of cruelty to animals understood in another way. If what happens to animals is a representation of what is happening to some humans then animal suffering must be staged to replicate human suffering, therefore there must be a belief that the animal can suffer in a way which is analogous to the human. Again, anthropocentrism and anthropomorphism are inseparable. Les Brown defines cruelty as 'unnecessary suffering knowingly inflicted on a sentient being.'[22] To be cruel in Brown's, as in Robert Bolton's, reading is to assert similarity. Where Alessandro Magno might be interpreted as celebrating his uncompromised human-ness as he emphasises the anthropoid qualities of the monkey, there is a possibility that the human-ness he is celebrating is inherently flawed. A human-ness based upon cruelty recognises the link with the animal and is not truly human at all. To enjoy cruelty there must be a recognition of suffering, but such a recognition implies sameness. The distinction of *anthropos* and anthropoid breaks down.

Raphael Holinshed's description of monkey-baiting makes a link between the ape and the spectator through its echo of Alessandro Magno's description of the monkey screaming, and it is a link which makes explicit the threat to the very sense of human superiority which might be thought to have been created in the Bear Garden. He writes,

> The like pastime also of the ... horsse with the ape on his backe, did greatlie please the people, who standing round, some in a ring upon the greene, other some aloft, and some below, had their eies full bent upon the present spectacle, diverse times expressing their inward conceived joy and delight with shrill sounds and varietie of gesture.[23]

The screaming monkey is no longer on horseback. With their 'shrill sounds and varietie of gesture' the humans themselves destroy the distinction between the watched animal and the watching human. Holinshed has placed the beast in the audience.

In a similar way Thomas Dekker's attack on the Bear Garden questions the assumption that the bears and the dogs are the only animals involved in the spectacle. Dekker begins by repeating Stubbes' and Perkins' gesture of reading the animal for its implications for human society:

> No sooner was I entred [the Bear Garden] but the very noyse of the place put me in mind of *Hel*: the beare (dragd to the stake) shewed like a black rugged soule, that was Damned and newly committed to the infernall *Charle*, the *Dogges* like so many *Diuels*, inflicting torments vpon it. But when I called to mind, that al their tugging together was but to make sport to the beholders, I held a better and not so damnable an opinion of their beastly doings: for the *Beares*, or the *Buls* fighting with the dogs, was a liuely represe[n]tation (me thought) of poore men going to lawe with the rich and mightie.[24]

The anthropocentric bias of writing about animals is unmissable: baiting is a reminder of human inequalities. But Dekker continues his description of the entertainment and fulfils the logic of anthropocentrism. He aligns some humans with animals:

> At length a blinde *Beare* was tyed to the stake, and in stead of baiting him with dogges, a company of creatures that had the

shapes of men, & faces of christians (being either Colliers, Carters, or watermen) tooke the office of Beadles vpon them, and whipt monsieur *Hunkes*, till the blood ran downe his old shoulders: It was some sport to see Innocence triumph ouer Tyranny, by beholding those vnnecessary tormentors go away w[ith] scratchd hands, or torne legs from a poore Beast, arm'd onely by nature to defend himselfe against *Violence*: yet me thought this whipping of the blinde *Beare*, moued as much pittie in my breast towards him, as y[e] leading of poore starued wretches to the whipping posts in *London* (when they had more neede to be releeued with foode) ought to moue the hearts of Cittizens, though it be the fashion now to laugh at the punishment.[25]

The colliers, carters and watermen fulfil Spiegel's logic of cruelty, they enact the role of the master, the beadle. But the cruelty to animals which takes place on the Bankside also destroys human status: the people whipping the bear may have 'the shapes of men', but they are merely 'creatures'. This is echoed in 1632 by Donald Lupton who wrote of the Bear Garden

This may better bee termed a foule Denne than a faire Garden. It's a pitty so good a piece of ground is no better imploied: Heere are cruell Beasts in it, and as badly us'd; heere are foule beasts come to it, and as bad or worse keepe it, they are fitter for a Wildernesse then a City ...[26]

The violence involved in taming wild nature – in expressing human superiority – destroys the difference between the species. The city, the place of humanity, is opposed to the wilderness, the place of savagery, but ironically it is the city which produces the very thing which makes its inhabitants more suited to the wilderness.[27]

The binaries of baiting and being baited; watching and performing; human and animal collapse into one another in dangerous and important ways. The only sense of being human voiced in these texts is one which is being destroyed. To note the suffering of the bear as Dekker does is, paradoxically, to diminish the sense of the separate nature of humanity. The Bear Garden emerges as a place of immense contradictions: the place which reveals the difference between the species also reveals their sameness. Baiting is the most explicit and spectacular site of anthropocentrism in the early modern period, but it is also the most explicit and spectacular site of humanity's confusion about itself.

More generally, to read for animals in early modern texts – as in the Bear Garden – is to find humans attempting to maintain their status. This status is so fragile that a general notion – humanity – can no longer function. Asserting human as a distinct category requires a limiting of the possibilities; requires exclusion. In human perception the problem of the status of animals can be dealt with only by making the dividing line between the species fixed, but in fixing the division it also becomes dangerously unstable. To separate man from beast is to assert that some humans are not human. This idea is not new. It can be traced back to Aristotle, and his division of master and slave is replicated in a number of areas of early modern society.[28]

III

The status of the native peoples of the New World was questioned by Europeans and their questioning led to such violent actions that in 1537 Paul III issued a Papal Bull stating that

> Indians were not to be considered 'dumb brutes created for our service'. Rather they were 'truly men ... capable of understanding the Catholic faith' ... that 'the said Indians and all other people who may later be discovered by Christians, are by no means to be deprived of their property, even though they may be outside the faith of Jesus Christ ... nor should they in any way be enslaved.'[29]

This Bull, however, did not seem to enter the imagination of many involved in the conquest and colonisation of the New World, particularly (and obviously) not those from Reformed nations. The brutalisation of the natives continued and '[w]hile no evidence of American bear-baitings has been found, references to Indian-baitings abound.'[30] For the colonists status is asserted explicitly through animalisation.

The animals closer to home, for Edmund Spenser at least, were in another colony, Ireland, and the problem of degeneration, of losing the qualities of human-ness, lie at the heart of *A viewe of the presente state of Irelande*. Not only are the Irish natives animalised – Irenius speaks of 'theire brutishnes and laothly filthines which is not to be named' – but the English settlers have, more dangerously still, 'degenerate[d] from theire firste natures as to growe wilde'.[31] Here the environment is blamed for the decline of the human: 'Lorde how quicklye dothe that country alter mens natures'.[32] In this context travel, relocation, becomes extremely dangerous, and is a danger which is experienced, as I discuss below, in London itself.

But it was not only the native other who was regarded as not human in early modern thought. In 1639 the activities of one New World slave owner were reported by an English visitor:

> *Mr Maverick* was desirous to have a breed of Negroes, and there-fore seeing [that his "Negroe Woman"] would not yield by per-suasions to company with a Negro young man he had in his house; he commanded him will'd she nill'd she to go to bed with her which was no sooner done but she kick'd him out again, this she took in high disdain beyond her slavery.

The inhumanity of the slave is further emphasised in 1655 by Henry Whistler of Barbados who noted that 'some planters will have thirty [slaves] more or lesse about four or five years ould: they sele them from one to the other as we doue shepe'.[33] Slaves are to be bred, not to be conversed with. By 1691 all inter-racial sexual relations were prohibited by law in Virginia. Echoing earlier interpretations of the monstrous births which were understood to follow human sexual relations with animals, children born of inter-racial relation-ships were termed 'that abominable mixture and spurious issue'.[34] These children were not human.

The animalised slave of the New World finds its parallel in the servant of England. When George Abbot, the Archbishop of Canterbury, accidentally killed one of the King's keepers while out hunting the King, wrote John Chamberlain, replied

> that such an accident might befall any man, that himselfe [had] the yll lucke once to kill the Kepers horse under him: and that his Quene in like sort killed him the best brache [bitch] ever he had, and therefore willed him not to discomfort himselfe.[35]

The servant, the horse and the bitch are the same thing to the King. Their deaths are equally insignificant.

Aristotle also aligned the slave – the domestic animal – with women and this alignment can be traced in early modern England.[36] In her advice manual Dorothy Leigh spells out the way to God to her sons, but she also includes some comments about good womanhood. She writes, 'the Woman that is infected by the sin of uncleannes, is worse the[n] a beast, because it desireth but for nature, and shee, to satisfie her corrupt lusts.' Chastity is the true quality of human-ness: 'let women be persuaded by this discourse, to embrace chastity, without which, we are meere beasts, and no women.'[37] This animality of women had an anatomical basis in Edward Topsell's description of the ape: following Aristotle, he states that 'The genitall or priuy place of the female [ape] is like a Womans, but the Males is like a dogges.'[38] The male human is like neither woman nor ape. It is merely like itself: splendid in its isolation.

In other writings the division of England into the court and the country is also frequently figured in terms of distance from and closeness to the beast. In a satirical dialogue between a Courtier and a Countryman Nicholas Breton presented the unbridgeable gap between the two societies. The 'gallant life of the Court', with all of its frivolous entertainments, is set against the simple rural ideals of 'speake well, and ride well, and shoote well, and bowle well'. Finally the Courtier tells the Countryman,

> I say this, that Nature is no botcher, and there is no washing of a black moore, except it bee from a little durty sweat: The Oxe will weare no Socks, howsoever his feete carry their favour: and Diogenes will be a Dog, though Alexander would give him a Kingdome: and therefore though you are my kinsman, I see it is more in name then in nature ...[39]

The animal will always be an animal, such is its nature, and the Countryman is like the moor, the ox and the dog. He is always not human, is always animal. The inclusion of the countryman in the status human is nominal rather than real.

It is not only the inhabitants of the country, though, who were criticised through alignment with animals. The court itself is also frequently represented as a place of brutalisation. John Taylor tells the story of a performing ape and his ape-ward travelling around the country with a warrant from the Queen.[40] The butt of the joke is apparently the rural audience, but the court also comes in for implicit criticism from Taylor. Whether the tale has any element of truth in it (and this has to be doubted), the publication of the narrative reveals the popularity of the idea of both provincial and courtly stupidity:

> Marry (said another senior [townsman]), wee see that by the Brooch in the mans hat that hee is the Queenes man, and who knows what power a knave may have in the Court to doe poore men wrong in the country? Let us goe and see the Ape, it is but two pence a peece, and no doubt but it will be well taken; and if it come to the Queenes eare, shee will thinke us kinde people that would shew so much duty to her Ape: what may she thinke wee would doe to her Beares if they came hither?
>
> .... This counsell passed currant, and all the whole drove of townsmen, with wives and children, went to see the Ape, who was sitting on a table with a chaine about his necke; to whom master Mayor (because it was the Queenes Ape) put off his hat, and made a leg; but Jacke [the ape] let him passe unregarded. But Mistris Mayoresse, comming next in her cleane linnen, held her hands before her belly, and, like a woman of good breeding, made a low curtsie, whilst Jack (still Court-like), although [he] respected not the man, yet to expresse his courtesie to his wife, hee put forth his paw towards her and made a mouth, which the woman perceiving, said: Husband, I doe thinke in my conscience that the Queenes Ape doth mocke mee. Whereat Jack made another mouth at her, which master Mayor espying, was very angry, saying: Sirrah, thou Ape, I doe see thy sauciness.[41]

The provincial citizens do not understand the reality of the Queen's warrant – that it does not have any real implications for the status of the performer – and they take the ape as the royal representative in their town and attempt to engage in human niceties with it.[42] Their stupidity sets them apart from Taylor himself and his knowing London readers, and they are brutalised: they curtsey to an ape.

However, in misrecognising the nature of the ape and regarding it as a representative of the court Taylor is also attacking the latter. The ape is mistaken for a courtier, and the figure of the court-ape is made literal. For the informed reader both the court and the country are aligned with the beast. Taylor's urban readers are the true humans in this text. But the process of brutalisation does not end here. In other texts the urban environment itself produces the means of losing humanity.

IV

It has been estimated that the population of London grew from 120,000 to 200,000 between 1550 and 1600, and the social upheaval which the growth of the city entailed also meant an increase in social alienation: '[a]lmost everyone was a migrant', and the 'conditions for a sense of isolation and insecurity' were in place.[43] In this context entertainment took on a very different function to the one it had in the provinces. The entertainments outside of the capital, such as the one Taylor mocks, were still very much a part of the festive communal calendar.[44] In Eccles bull-baiting was part of the September Wake, and at Stone in Staffordshire the patronal festival of St Michael and All Angels was celebrated with bull- and bear-baiting and dog-fights.[45] The church in Winterslow, Wiltshire actually paid for cock-fights and throwing-at-cocks at Easter throughout the period 1540 to 1640.[46]

In London, however, although the links with festivals such as Shrove Tuesday, Ascension Day, Mayday, Midsummer and Saint Bartholomew's Day did continue, 'the impression remains that ... traditional festivals were in relative decline in London'. Peter Burke has written:

> In a town which was growing from about 170,000 to about 550,000 people [in the seventeenth century], it was worth the while of a variety of professional entertainers – acrobats, actors, ballad-singers, bear-wards, clowns, fencers, puppet-showmen – to put on a virtually continuous performance. Where villagers might see this kind of show a few times a year, Londoners could see them all the time.[47]

Entertainment was removed from the civic, religious and communal calendar and the commercial spectacle became an end in itself.

For many commentators the gap between diversion and damnation in the capital was perceived to be a small one.[48] A list of the abuses of the age voiced by 'St. Pavles-Chvrch' in 1621 shows a wide range of recreational choices open to the citizen of London:

> To see a strange out-landish Fowle,
> A quaint Baboon, an Ape, an Owle,
> A dancing Beare, a Gyants bone,
> A foolish Ingin moue alone,
> A Morris-dance, a Puppit play,
> Mad *Tom* to sing a Roundelay,
> A Woman dancing on a Rope;
> Bull-baiting also at the *Hope*;
> A Rimers Iests, a Iuglers cheats,
> A Tumbler shewing cunning feats,
> Or Players acting on a Stage,
> There goes the bounty of our Age:
> But vnto any pious motion,
> There's little coine, and lesse deuotion.
>
> For euery fashion base and vaine,
> For purchasing, or greedy gaine,
> For Dicing, Drinking, foolish sporting,
> Hunting, Wenching, Coaching, Courting,
> There is enough in euery Function,
> But to this Church is small Compunction.[49]

The writer bewails the church's loss of power and the destructive seductions of entertainment.

Keith Wrightson has described the importance of community ties in rural England during the early modern period as a form of 'neighbourliness': 'a horizontal relationship, one which implied a degree of equality and mutuality between partners to the relationship, irrespective of distinctions of wealth or social standing'.[50] At the heart of this community was the church. But, as St. Paul's itself noted in Farley's poem, this sense of community was left behind when people moved to the capital, and in its place was entertainment.[51]

The commercialisation of entertainment is clearly linked to the growing economy.[52] Pleasure was no longer provided by the civic government or the church, it was paid for by the individual at a time of his or her own choosing.[53] Viewing became an isolating experience which was organised, like so many other things, along class lines. The horizontal relationship which Wrightson sees in rural neighbourliness is, in the city, replaced by vertical relations.[54] As in Spenser's Ireland, a move to the city changed the nature of humanity itself.

Judgement, the thing which Taylor's rural audience lacked, comes at a price in London. In 'THE INDVCTION' to *Bartholomew Fair* Jonson famously noted:

> it is further agreed that every person here haue his or their free-will of censure, to like or dislike at their owne charge, the *Author* hauing now departed with his right: It shall bee lawfull for any man to iudge his six pen'orth, his twelve pen'orth, so to his eigh-teene pence, 2. shillings, halfe a crowne, to the value of his place: Prouided alwaies his place get not aboue his wit. And if he pay for halfe a dozen, hee may censure for all them too, so that he will vndertake that they shall bee silent.[55]

Poor judgement is literally the judgement of the poor. Wealth becomes a signifier of more than merely social status; it signifies the ability to understand. But in London new money never lasted long, it brought in its wake new temptations which destroy judgement which is itself aligned by John Marston with human-ness. Marston figures the gallant spendthrift as worse than 'old Iack of Parris-garden':

> Euen Apes & beasts would blush with natiue shame,
> And thinke it foule dishonour to their name,
> Their beastly name, to imitate such sin
> As our lewd youths doe boast and glory in.[56]

In exercising judgement – spending money – the gallants reveal themselves to be animals. Marocco, the famous intelligent horse, voiced a typical complaint about the dangers of the capital in his dialogue with his owner Bankes: 'He that will thrust his necke into the yoke, is worthy to be used like a jade. He that hath been a gentle man of fair demeanes and will so demeane him selfe to let landes and lordeshippes flie for a little bravery ... let him crye'.[57] The power which wealth brings is tempered by the danger which also follows: like the Bear Garden, money seems to make humans truly human, but in fact it achieves the reverse.

The sense of London as both centre of wealth and place of moral insecurity can be traced in many contemporary sources. In 1606 Thomas Dekker wrote,

> O *London*, thou art great in glory, and enuied for thy greatnes: thy Towers, thy Temples, and thy Pinnacles stand vpon thy head like borders of fine gold, thy waters like frindges of siluer hang at the hemmes of thy garments. Thou art the goodliest of thy neighbors, but the prowdest; the welthiest, but the most wanton. Thou hast all things in thee to make thee fairest, and all things in thee to make thee foulest ...[58]

This dichotomy fairest and foulest (paralleled, of course, in *Macbeth* which was probably also performed in 1606)[59] was repeated throughout the period, with the attractions of the capital offset by the corruption to which they would inevitably lead. In 1628 Richard Rawlidge – this time in language reminiscent of John of Gaunt's eulogy to the lost England in Shakespeare's *Richard II* – affirmed the double-edged nature of London's greatness;

> *London* is not alone the Honour of our Nation, but by others reputed the mirror of *Europe*. ... But (with griefe of heart I speake it, and the reproch of all those who haue caused or suffered it:) this so renowned, so famous a Place, this peerelesse Citty, this *London*, hath within the memory of man lost much of hir pristine lustre, and renowne ...[60]

In response to the temptations available in the capital the advice was clear: 'And (my sonne)', warns Richard Johnson, 'bee thou thus conceited, that the man that is enticed to bee a Dicer, of his owne accord will become a Whore-maister'.[61] The implications were that dissolute pleasures would lead to a descent into the underclass. Thomas Adams makes the step even more dangerous:

> If euer they [i.e. gallants] begin any worke with the day, they dispose of it in this fashion; First, they visit the Tauerne, then the Ordinary, then the Theater, and end in the Stewes: from Wine to Ryot, from that to Playes, from them to Harlots.

.... Here is a day spent in an excellent methode: If they were
Beasts, they could not better sensualize, it would be but lost
labour to tell them ...[62]

The pleasures of the capital create a climate in which the human is
turned into an animal. What is essential is also changeable.

For many of the moralisers in early modern England those who
succumbed to temptation fell into the category of the animal, and
the idea of the beast which these writers invoke has, as Levinas
showed, dangerous implications. Anthony Pagden has argued that
'[d]ehumanisation is, perhaps, the simplest method of dealing with
all that is culturally unfamiliar'.[63] Those who are strange, different,
or other can be exploited because they are represented as not
human. Conversely, the representation of certain groups as not
human must also be seen as a result of their exploitation. If you
abuse something or someone you need a language through which
to represent it as legitimate treatment rather than abuse, and at its
most potent the language of brutalisation is internalised by the
social groups who are labelled as animal.[64] Breton's Countryman, in
response to the Courtier's accusation that he is of a different nature,
states 'better be a mannish Dogge then a dogged man.'[65] The
urban(e) writer can represent the animal nature of the rural population of England as an accepted part of their existence; the country-
folk acquiesce in it.

The identification of the self as animal makes an interpretation of
baiting which rests on the always-alreadiness of human status
impossible. Baiting makes the human an animal. As well as this,
however, the groups within society who are animalised are merely
anthropoid like the monkey on horseback, and not *anthropos*:
mannish-dogs rather than dogged-men. With a change in the
perception of the spectators comes a change in the perception of
the entertainment.

It might be argued that the constant bestialisation of humanity in
early modern England makes it a cliché rather than an important
social construction. But what is evident is that the animal is a pre-
occupation, and that there is no constant sense of the human to be
found. Fundamentally, who is human and who is not is never clear,
some are human in one place and not human in another.
Ultimately, different groups occupy the position of the animal in
order to affirm the human status of others.

V

The dissolution of the human can be traced in two emblems from either end of the period covered in this book. The change in the emblems reveals the increasing significance of the debate which I am interested in. The woodcut in Figure 1 comes from Geffrey Whitney's *A Choice of Emblemes and Other Devices* (1586) and

Fig. 1.   'In curiosos', from Geffrey Whitney, *A Choice of Emblemes and Other Devices* (1586).

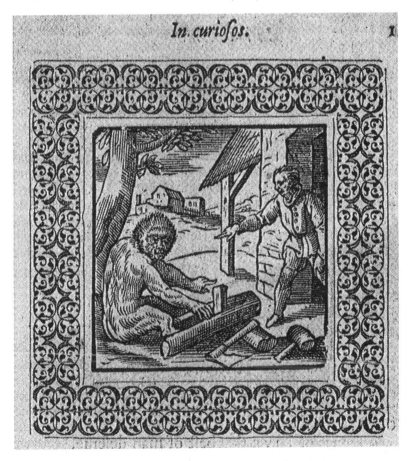

accompanies the traditional tale of the imitative ape. The second verse reads

> Let none presume an others arte to vse,
> But trie the trade, to which he hath bene kept:
> But those that like a skill vnknowne to choose,
> Let them behoulde: while that the workeman slept,
>   The toying ape, was tempringe with his blockes,
>   Vntill his foote was crush'd within the stockes.[66]

The ape, in typical fashion, imitates humanity and fails.[67] The attempt to overstep natural status figured in the image is paralleled in verse by the attempt to overstep social status. The message of Whitney's emblem is apparently a conservative one: social order – the clear and limiting structure of early modern society – is in place to protect the members of that society. Knowing your own place is about knowing when you are well-off and overstepping your place is about ignorance, animality.

The woodcut, however, also does something slightly different. The workman points at the ape just as Whitney points at the apishness of social climbing. But the ape itself – huge and muscular – looks out of the frame at the reader and itself gestures towards the workman. This is not the silent animal of metaphor or symbolism, the animal who only ever speaks with a human voice, and appears, as Beryl Rowland has argued, with a human face.[68] This is the not-quite-but-almost human who offers a threat. The human and the ape of Whitney's woodcut are both readers: the man examines the ape, the ape examines us. The perceiving animals of the title of this book are the readers who were expected to distinguish the beast from the human, and are the animals who were used to examine human culture.

Nearly fifty years later, however, the emblem of imitation has changed. Instead of the ape we now find the dwarf, the '*Apish-Pigmie*' of George Wither's *A Collection of Emblems Ancient and Modern*.[69] The most famous dwarf of the 1630s was the Queen's servant Jeffrey Hudson who was regarded by the court as the '*most perfect abridgement of Nature*'.[70] However, this perfect abridgement was the son of the man who 'kept and ordered the baiting Bulls for George Duke of Buckingham' and was introduced to Queen Mary when he was served up to her in 'a cold baked Pye'.[71] Hudson's qualifications for becoming an animal were impeccable.

Whether or not this is a direct reference to Hudson, and no such link can be made categorically, Wither's emblem plays on the animality of the dwarf, and its *subscriptio* is analogous to Whitney's ideal of social stability:

> A *Pigmey-spirit*, and an *Earthly-Minde*,
> Whose looke is onely fixt on Objects vaine;
> In my esteeme, so meane a place doth finde,
> That ev'ry such a one, I much refraine.[72]

The dwarf does not look to the greater vision beyond the world but looks only at itself. He does not perceive the heavens and fails to achieve the status of a true human. Thomas Morton, following Plato, said that 'the brute beastes ... never once lift vp their eies to heauen'.[73] In the image (Fig. 2) the dwarf is admiring his new-found and false stature. The fragile stilts which raise him to the height of the human parallel the wood blocks which allow the ape, for an instant, to replicate the workman. The ape traps its foot; the dwarf, we are asked to acknowledge, is bound also to fall.

But there is a new understanding of the implications of the loss of status in Wither's emblem. In the first image we perceive the animal and it perceives us. In the second the dwarf can only look within and regard its own lost status. The human world of riding a horse continues in the background, but the dwarf, static in his self-obsession, is separated by a dark and foreboding area of rock. To make the return to the world pictured in the background the dwarf must abandon the stilts, and in doing so he will abandon his false status, his imitation of humanity. The dwarf's attempt to be human can never work. The stilts give him the stature he needs – the quality of human-ness – but they also reveal his stature to be false. He is never a natural human.

But the humans in the background also have a false stature; are also raised by unnatural means. In place of stilts there are horses, or there is a carriage. The height which the dwarf aspires to, which is recognised as human, is revealed to be ideal and not real. The height of the human exists only through using an animal. To ride a horse is to gain status, but it is the equivalent of the dwarf's fragile stilts. In the emblem the figures on the ground would seem to occupy the natural level, the truly human state, but pedestrianism brings with it another alignment, this time with the dog. Wherever they stand, whatever their height, the animal is always present as a marker of human status.

Fig. 2.   'Illvstr. XIIII', from George Wither, *A Collection of
Emblems Ancient and Modern* (1635).

ILLVSTR. XIIII.   Book. I.

   Both Whitney's and Wither's emblems underline the problem of
being human in early modern England. In one an animal imitates
the human and fails, and in its failure is threatening rather than
comforting. In the other the dwarf recognises his loss of status and
attempts to reassert himself. In his attempt, however, a link
between stilts and horses emerges which underlines the false
nature of the human-ness he aspires to. The rest of this book traces
the attempts to establish a fixed sense of human-ness in early
modern culture. Moving through religious, humanist, scientific, and
legal debates, what is uncovered is a sense of the animality of
humanity and the inevitable and paradoxical exploitation of

animals. But the final chapter offers a new sense of human-ness. Richard Overton argues that human status can be achieved, but only by discounting the false logic of previous models and accepting what might be termed 'pedestrian' status. Difference can only be achieved by dismounting, by walking on a level with the dog.

2

Wild Beasts Making Havoc of the Soul: Animals, Humans and Religion

William Perkins offers what seems like a very straightforward definition of the difference between the human and the animal:

> the proper subjects of co[n]science are reasonable creatures, that is men and Angels. Hereby conscience is excluded ... from bruit beasts: for though they haue life & sense, and in many things some shadowes of reason, yet because they want true reason, they want conscience also.[1]

This interpretation of species difference erects a firm and clear boundary between human and animal within religious ideas: the human is a being with a conscience, the animal lacks one. Reformed theology uses this as a basis for discussions of the human relation to the animal world in many different ways, but ultimately the result is always the same. Human dominion follows the animal's lack of conscience.

This chapter will argue, however, that an examination of the early modern understanding of the child, the atheist, the werewolf and the wildman undermines the distinction which is based upon the possession of conscience. These humans all venture dangerously close to the Reformed notion of the animal and question Perkins' proposal. Respectively, if conscience is based on understanding and an infant is acknowledged to have none, then its possession of a conscience, and therefore its human-ness, must come under question. If we are predestined to either salvation or reprobation, as Reformed theology argues, then the atheist is either making a choice – an impossibility within predestinarian ideas – or is made by God, is always-already an unbeliever, an animal. The werewolf is a logical extension of the doggish-unbeliever and in its acts of violence it likewise reveals two dangerous possibilities for a descent from the human into the animal: either of the loss of the conscience, with the implications that the human can slip into the animal; or of violence – wildness – as a part of the human conscience, a possibility which once again questions the status of human. Neither fits comfortably within Reformed theology. Finally, the wildman appears. This is a figure without access to the education which is so important in debates about infants, and what emerges in the changes which occur to the archetypal tale of *Valentine and Orson* between 1565 and 1637 is the sense in which Reformed understandings of the human have had to be abandoned. The human is no longer figured as a creature of conscience, the use of this as a differentiating feature having fallen apart; instead it is a creature of civilisation. The original site of difference – the assertion of the animal's lack of conscience – is shown to be too frail to work: there are too many humans who seem to lack a conscience, who are, implicitly, animal.

This chapter concentrates in large part on the work of William Perkins which represents a specifically Calvinist interpretation of Reformation ideas.[2] It is these ideas which I am interested in here because Perkins was perhaps 'the most widely-read and influential puritan theologian' of his age.[3] He was, in his day – the late Elizabethan period – famous for preaching 'the terrors of the law and was reputed to be able to make his hearers' hair stand up and their hearts fall down by his pronunciation of the word "damn".'[4] The differences and debates within Reformed ideas in England in the late sixteenth and early seventeenth century are not dealt with here, rather the chapter offers a picture of the way in which one important and influential interpretation of Reformation theology forced a reappraisal of the status of humanity. I start with the new Reformed human, I move later to its deformity.

I

An important image of prelapsarian humanity which existed before the Reformation can be traced in the ideas of Giovanni Pico della Mirandola. In *On the Dignity of Man* he represented God saying to Adam

I have placed thee at the center of the world, that from there thou mayest more conveniently look around and see whatsoever is in the world. Neither heavenly nor earthly, neither mortal nor immortal have We made thee. Thou, like a judge appointed for being honorable, art the molder and maker of thyself; thou mayest sculpt thyself into whatever shape thou dost prefer. Thou canst grow downward into the lower natures which are brutes. Thou canst again grow upward from thy soul's reason into the higher natures which are divine.[5]

The descent into the brute is original man's to contemplate, and to avoid. God leaves the human will all powerful. Degeneration is a God-given, but human, choice. Such freedom of choice is also offered to Calvin's Adam, but the choice he makes reduces him to the status of the beast: 'it is impossible to think of our primeval dignity without being immediately reminded of the sad spectacle of our ignominy and corruption, ever since we fell from our original in the person of our first parent.'[6] The Fall which depraved Adam and his posterity was within Calvinist thought irreparable by humanity acting alone. The dignity of Adam which Pico proclaimed was transformed into Calvin's sense of humanity's 'wretchedness'.[7] There was no sense, as there was within Pelagian ideas, that 'humans are able to merit their salvation', that good works can lead to heaven.[8] Instead there was the belief that salvation came from God alone, from the gift of grace. The function of the Reformed human was 'not to choose but to be chosen.'[9] There was a movement in Reformed ideas away from Pico della Mirandola's man of dignity towards Calvin's vision of humanity: 'all, without exception, are originally depraved.'[10]

Calvin's understanding of human weakness was taken up by William Perkins. In *A Treatise of Vocations or Callings of Men* Perkins notes 'the heathenish opinion of men, which thinke that the particular condition and state of man in this life comes by chance: or by the bare will & pleasure of the man himself.'[11] It is not luck or human effort which gains salvation. Perkins is presenting (in the negative) two ideas which were central to Reformed thinking: providentialism and predestinarianism.

The providential design had important implications for humanity. Calvin's understanding of the organisation of the universe was simple; 'All future events being uncertain to us, seem in suspense as if ready to take either direction'; they *seem* to be, but this is misperception because in reality 'nothing will happen which the Lord has not provided'.[12] Calvin's God was not a creator who 'completed his work once and for all, and then left it.' He was 'by a special Providence sustaining, cherishing, superintending, all the things which he has made, to the very minutest, even to a sparrow.'[13] Calvin's ideas were reiterated in England by Thomas Morton: God's 'permission is not idle, but effectuall and working: for God is not an idle looker on, as if he had cast off the care of the world and of his creatures, and left all at sixe and seauen (as we vse to speake) but he hath a parte in this worke'.[14]

In place of the power to choose which was offered to the first man, Reformed thinkers proposed absolute powerlessness for fallen humanity. Predestination meant that the salvation or reprobation of the individual Christian was not in their own hands but was already decided. Pico's vision of Adam represents the abandoned status for Calvin. Postlapsarian Adam is a very different being; and the difference was transmitted to all of his posterity. For Calvin individuals had to accept their inability to alter their eternal fate: God was the fair, true, unbiased, and unquestionable judge of all. There was no sense of man as deserving of salvation, quite the opposite: Calvin wrote that 'when men are judged by their natural endowments, not an iota of good will be found from the crown of the head to the sole of the foot'.[15]

In emphasising predestination Calvin and Reformed theologians such as William Perkins proclaimed a new vision of humanity. Perkins wrote that 'All men are wholly corrupted with sin through *Adams* fall, and so are become slaues of *Satan* and guilty of eternall damnation.'[16] The Reformed human lacked dignity, just as he lacked, on his own, immortality. To be immortal was a gift from God; predetermined, not dependent upon human effort or work. The power to enact salvation for the self – the Pelagian position – had gone, and what remained was the Augustinian position: the powerless human.

But at the heart of human frailty there was still one site of human power. In the creation Adam was given absolute rule over the animals. John Moore termed prelapsarian man 'a petty God ... all things being put in subjection under his feete'.[17] But the eating of the forbidden fruit changed everything: Adam and Eve's sin destroyed their petty-Godhood; 'if man had not sinned, no creature had been hurtfull to him', noted Gervase Babington the Bishop of Llandaff.[18] Thomas Morton wrote that postlapsarian man 'is not now Lord of the creatures; for they do oftener tyrannise ouer him, deuouring him & tearing him in peeces, the[n] perform any seruice

vnto him'.[19] The destruction of the perfection of Eden and the wildness of animals were used by theologians to remind humanity of
their fallen state. Perkins wrote that 'when we see any creature that
is hurtfull and noisome vnto man, and would rather deuoure then
obey him; it must put vs in mind of our sinne'.[20] In the same vein
Thomas Draxe wrote

> Let us in no wise curse, banne, blame or misuse any of the poore
> Creatures, knowing that if there bee any defect or untowardnes
> in their nature; or any want of duty and observance in them,
> towards us, our sinne hath been and is the cause and occasion of
> it ...[21]

Calvin wrote that Adam lost the 'superiority & rule which God had
giuen vnto him.' Where the human is bestial in the face of God, the
animal becomes a threat to humanity: 'But so soone as he began to
be obstinate and rebellious against God, he felt the fiercenes of the
brute beastes against him.'[22] Human wildness led to animal wildness. After the Fall humanity's power over the animal was enforced
rather than agreed.

Despite the dangers of animals, however, an order which situated
humanity at the top of the natural scale was still in place.[23] In 1631
Edward Cooke proposed (in rather awful verse) that the 'feare of
Man which keeps [animals] thus in awe, / Was through *Gods* mercy
giv'n them as a law'.[24] It is only out of pity to humanity that any
rights over the animals are maintained. But it is not only the issue of
rights which are important here. The question of the nature of the
animal is also significant because in representations of animal
nature are found discussions of human nature.

There were some very important differences between Reformed
and Thomist representations of animals. William Perkins proposed
that animals were 'shadowes' of the human, poor and insubstantial
reflections.[25] In *Summa Contra Gentiles*, on the other hand, Aquinas
argued that the animal was all body and no soul, a truly substantial
figure.[26] However, despite the differences in the understanding of
animal nature, there was some agreement between Thomist and
Reformed belief on the question of the treatment of animals. In the
thirteenth century Thomas Aquinas wrote that animals could be
'loved from charity as good things we wish others to have', that a
person who pities an animal is more likely to pity a human being.[27]

In the early seventeenth century this view was still dominant when John Rawlinson summarised it as 'Save a beast's life and save a mans'.[28]

Rawlinson's sermon, *Mercy to a Beast*, begins with Proverbs 10: 12 – 'A righteous man regardeth the life of his beast' – which is initially read to mean that a '*bond of Oeconomicall* or *houshold righteousness*' should exist between a man and his family and between a man and his beasts. Regard for animals comes in six 'braunches': feeding; sparing from overwork; pardoning infirmities; feeling compassion for; ruling and guiding; and protecting and defending. As he notes, 'If our beasts regard *us*, because we are *their masters*; ought not *we* then to reguard *our* beasts, because they are *our servants*?'[29] However, to this sense of human duty Rawlinson adds the anthropocentric belief that regarding the life of the beast is really a call to regard other men; 'that *reguard*, which a righteous man hath, as of his *beast*, so much more among *men*'.[30] The generosity towards animals, as in Aquinas, is not for its own sake: in our kindness we recognise the importance of humanity. Animals are a means of displaying benevolence which is in reality to be exercised on humans, and is, ultimately, self-serving:

> For if thou pittie not, rather than punish the infirmities of thy beast, which through thine owne sinne is become lesse plyant and tractable, than otherwise it was, and would have beene; what do'st thou else, but teach God not to pardon, but punish thine offenses, who by thy sin, are now degenerated into a *beast*.[31]

The animal in Rawlinson's sermon becomes, in typical fashion, the vehicle for human thoughts about the self. We are kind to animals in order to be kind to other humans, and the distinction between the two categories is vital. To be kind to animals because they deserve kindness in and of themselves is almost unthinkable. As Thomas Adams noted, 'if they will preferre beasts before men, let their portion be among the beasts.'[32]

This sense of the overriding importance of humans is reflected in the absence of animals in heaven. One of the few early modern writers to tackle the issue of the immortality of the animal soul was the otherwise orthodox Thomas Draxe.[33] In *The Earnest of our Inheritance* not only does Draxe, like Perkins, ask the Christian to look within when animals are cruel, he also proposes that animals (and plants) will be found in heaven. This potentially heretical statement needs to placed in the context of Romans 8: 22–3:

> For we know that the whole creation groaneth and travaileth in pain together until now. / And not only they, but ourselves also, which have the firstfruits of the Spirit, even we ourselves groan within ourselves, waiting for the adoption, to wit, the redemption of our body.

Draxe interprets this to mean that, excluding hybrids, mongrels and monsters, all shall take part in the Resurrection. This apparent endowing of the whole of the natural world with an immortal soul is not, however, what it seems. Draxe continues,

> though [plants' and animals'] naturall ends, and uses which served for mans infirmitie shall cease; yet their spirituall and principall ends, to witt, the setting forth of Gods glorie: the matter of mans delight, and the exercise of his meditation and thankfulnesse, may and shall continue and abide.[34]

Animals are in heaven to fulfil a function; they are not resurrected because they are regarded as sentient beings in and of themselves. The animals begin and remain as symbols of God's power. They are reminders, even in the afterlife, of the centrality of humanity.

The sense of the absolute difference of the animal and the human seems firmly in place. The animal can be dominated or represent human power, it has no power or potential of its own. But this stability only works when reading animals; when we look at the humans enacting dominance we get a blurring of the sense of difference. While the animal is firmly animal it is the human who constantly fails to be truly human. To begin to trace this failure we need to go back to the beginning of human life and to the status of baptism in Reformed ideas.

II

The Catholic belief that a child was incapable of mortal sin disappeared in Reformed thought.[35] In its place was Calvin's belief that

> even infants bring their condemnation with them from their mother's womb; for although they have not yet brought forth the

fruits of their unrighteousness, they have its seed included in them. Nay, their whole nature is, as it were, a seed of sin, and, therefore, cannot but be odious and abominable to God.[36]

This sense of the child as always-already sinful was taken up by a number of writers: William Gouge, for one, argued that 'Children drew contagion from their parents'.[37] In a defence of the child, however, William Hubbocke revealed the awful implications of Calvin's belief:

> O pitifull & cruell sentence, whose eares will not tingle at it? Infants who cannot speak, think, or do ill, the child whose flesh is scarce curded in the wombe; whose bones are scarce gristled out of the wombe: from the darknes of the wombe passe to the vtter darknesse for euer.[38]

The question, how can the child – the truly incapable – have sinned? is answered with recourse to the Fall. The child is the child of Adam, and therefore the inheritor of Adam's sinfulness. There is no avoiding this. But, as the baptism service in *The Booke of Common Prayer* states, there is a possibility of change: 'O mercifull God, graunt that the olde Adam in these children may be so buryed, that the newe man may be raysed up in them.'[39]

Initially inspired by Martin Luther's return to Augustine, Reformed thinkers adhered to the belief in justification by faith alone – the belief that righteousness and salvation come only from God's giving of grace. Perkins wrote, 'Faith is a wonderfull grace of God, by which a man doth apprehend and applie Christ, and all his benefits vnto himselfe.'[40] Such justifying faith was necessary for salvation, without it there was only reprobation, the eternal loss of God in Hell. But within Reformed thought existed the notion of 'double justification'.[41] The initial and vital justification of the Christian came from God in the form of grace and was invisible, but the second justification -- termed regeneration – was visible, apparent in the moral actions of the Christian. Good works could be performed in order to *prove* salvation rather than achieve it and within Reformed ideas regeneration often took the form of education and, apparently at least, baptism.

The sense of baptism as the place of renewal raises difficulties within Reformed logic. Baptism could not transform the infant because this would be dangerously Pelagian, would make it an act of human choice, one made by the parents and the priest. Instead it must merely symbolise the movement from old to new Adam: as was noted in *The Second Tome of Homilies*, a sacrament is 'a visible signe of an inuisible grace, that is to saye, that setteth out to the eyes and other outwarde senses, the inwarde workyng of Gods free mercye'.[42] An examination of Luther's theology of baptism – the most influential in England – reveals the complex ways in which this mercy worked.[43]

Luther wrote, '[w]henever we hear the word and are baptised, there we enter into eternal life.' The implications of this, Karl Barth argues, seem to be that 'a small child becomes a Christian in baptism'.[44] The act of baptism becomes problematically tied up with the act of gaining justification which is, for Luther and other Reformed thinkers, the act of God alone. In the providential order of the Reformed world it was not for humans to create themselves, but for the divine; it was not for the individual Christian to seek for grace, but for that grace to be given by God. For Luther, however, baptism is a key representation of the work of God. It is 'an appointed place for mankind's encounter with the divine'.[45]

In opposition to the condemnation of outward signs of worship (found in the Elizabethan vestiarian controversy, for example), Luther's understanding of baptism was that it was chosen *by God* for his meeting with humanity, it was not made by man.[46] Luther argues that baptism is a symbol of humanity's salvation which can only work if faith (and therefore grace) has been conferred *before* the baptism takes place. While a child may be naturally damned (as all of Adam's posterity are) this does not mean that s/he is automatically reprobate. As Kendall notes, damnation is different from reprobation: 'all the reprobate are born into the state of damnation, not all who are born into the state of damnation are reprobate.'[47] This interpretation of baptism can be seen in the case of Nathaniel the Jew, who was baptised in London on 1 April 1577. It was paramount that conversion precede baptism; a Jew could never be christened.[48] Within Reformed ideas – Nathaniel's confession of faith is reproduced with a sermon by the Protestant martyrologist John Foxe – Nathaniel must always-already have been saved, and the baptism, like his confession of faith, is merely a symbol of his salvation. It does not enact salvation.

So Luther's ideas offer a sense in which even before baptism the infant has faith, otherwise the implications would be that faith is a product of baptism (and therefore humanity) itself. To use Trigg's formulation, Luther proposed not that baptism produced faith, but that baptism 'adorns the baptised with holiness; with righteous-ness; and even with wisdom.' The faith of the infant (*fides infan-tium*), or any convertee, represents the always-alreadiness of faith in the saved human. Faith is given by God, and is always present, not in its fully fledged state but 'as a divine possibility and creation in man, as opposed to a natural human possibility or work.'[49] The lack of the fully fledged state makes the Reformed theory of double justification seem unavoidable. But where faith is divine all other routes – education, good works – are purely human means of achieving salvation. Alone they mean nothing, they represent the Pelagian position, but with faith they can mean everything. The importance of a Godly education in early modern ideas is well documented, and it must be situated in the context of the idea of regeneration.[50] But ultimately regeneration becomes problemati-cally tied up with the status and capacity of humans.[51]

The educational duties of the parent to the child are well known in the early modern period. William Perkins proposed that the parents should have their child baptised and give it a name and that they should 'endeauour to sowe the seedes of godlinesse and religion in the heart of the child, so soone as it comes to the vse of reason and vnderstanding'.[52] Similarly, in their reading of the fifth commandment, John Dod and Robert Cleaver emphasised the role of education:

> The parents dutie to the children in their tender yeares and child-hood, is first to instruct them in religion, to season them with the words of pietie, or by little and little to drop in the grounds of holiness into them, even so soone as they are able to speake and beginne to haue the least vse of vnderstanding.[53]

Baptism, however, had a particular impact on the issue of duty to the child. William Coster's formulation that 'baptism was a second birth, the priest a spiritual midwife, and the profane, natural parents replaced with spiritual ones' is a useful one.[54] In the Christening of the child – in the change of the old Adam into the new – a new form of guidance was introduced. In *The Booke of Common Prayer* the role of the Godparent is spelt out in some detail, and it is worth citing the section of the baptism service at length here:

> Forasmuch as these chyldren haue promised by you to forsake the deuyll and all his workes, to beleue in God, and to serue hym:

You must remember that it is your partes and duties to se that these Infantes be taught, so sone as they shalbe able to learne, what a solempne vowe, promise, and profession they haue made by you. And that they maye knowe these thynges the better, ye shall call vpon them to heare Sermons. And chieflye ye shall prouide that they may learne the Crede, the Lordes prayer, and the ten Commaundements in the Englyshe tongue, and all other thinges which a christian man ought to knowe & beleue to his Soules health ...[55]

These are the bare minimum requirements for the good Christian. In her counsel to her sons Dorothy Leigh left similarly clear instructions on the duties of a Godparent:

if any shall at any time desire you to bee a witnesse to the baptizing of their childe ... you shall desire the person so desiring, to giue you his faithfull word, that the child shal be taught to reade, so soone as it can conueniently learne, and that it shall so continue, till it can read the Bible.[56]

Her emphasis on literacy rather than just on *learning* is, of course, limited in its application; even in 1642, 70 per cent of adult men were illiterate.[57] It it is for these reasons – the need for a knowledge of the faith, and a lack of literacy – that the catechism was becoming increasingly important in English religious ideas. A massive number of catechisms were published in the late sixteenth and early seventeenth centuries, and their popularity reveals the importance of achieving a true understanding of Christian ideas without the need for, or possibility of literacy.[58] Henry Holland defined catechising as 'a forme of instruction, wherein the same matter is often repeated that the weake may the better conceiue it, and remember it.'[59] In *The Foundation of the Christian Religion Gathered into Six Principles* Perkins presented a collection of beliefs to be learned by 'ALL IGNORANT PEOPLE *that desire to be instructed'*.[60] The ignorant person would learn by heart six basic ideas. The 'second principle' is a useful illustration here of the catechitical method:

Q. What doest thou beleeue concerning man, and concerning thine own self?

A. All men are wholly corrupted with sin through *Adams* fall, and so are become slaues of *Satan*, and guilty of eternall damnation.[61]

The exposition, or more complex reading of this principle, spells out the Reformed idea of the status of humanity even more clearly:

Q. Let vs now come to our selues, and first tell mee what is the naturall estate of man?

A. Euery man is by nature *dead* in sin as a loahtsome [*sic*] carrion, or as a dead corps lieth rotting and stinking in the graue, hauing in him the seed of all sinnes.[62]

The repetition of this understanding of humanity which is called for in Perkins' catechism works to fulfil, through the words themselves, the Reformed sense that human power and immortality are found, paradoxically, through the acknowledgement of human powerlessness. In this way Reformed thinkers present learning as part of the process of regeneration. The catechism offers the child the education which is promised in baptism, and it teaches a clearly Reformed line on the status of the individual Christian. William Hubbocke proposed that 'the infant cannot reason, yet it hath the seed of reason'; it is the process of education which will – to extend the metaphor – water this seed.[63] Humans are, Reformed beliefs argue, utterly destitute but utterly able to learn. Salvation does not cancel out sinfulness, God's mercy is all the greater because of the undeserving nature of humanity. We are saved even though we are damned.

Because of the sinfulness which is in the heart of all Christians, however, learning is also vital; even the illiterate must have access to the truths of the Bible. Repetition, however – what is called for in catechisms – may seem to be very different from understanding, and in this age of printing and increased emphasis on the interpretation of the scripture, the problem of the status of the illiterate masses emerges. Perkins' emphasis on the *desire* for instruction is one way of dealing with this issue: the ignorant believer is a true believer in their *will* to learn. It is implicit that such a will is God-given, and therefore in itself proof of regeneration and ultimately salvation. Will is, in fact, to be found in the operation of conscience. When an infant is baptised a promise is made by the new parents that the child will be educated. This education does not displace grace in the salvation of the child, it merely develops a quality which God has placed as potential in the infant. Education may be difficult, a labour, but it is also predestined. Baptism is the symbolic beginning of the life of the Christian, the end of which is the full and sanctified operation of conscience, and it is in the possession of a conscience that the human is distinguished from the animal.

III

For William Perkins the conscience is 'a part of the vnderstanding in al reasonable creatures' which reveals the invisible workings of God: the true believer looks not for external symbols of salvation but for internal proofs.[64] It is not only a poetic technique which made Sir Philip Sidney's muse advise him to 'look in thy heart, and write', but a product of Protestantism: the truth lies within the individual Christian, not outside him or her in the symbols of the external world.[65]

Perkins' model of the Christian understanding is a tripartite structure of mind, memory and conscience. The mind is the 'storehouse and keeper of all manner of rules and principles'; memory 'serues to bring to minde the particular actions which a man hath done or not done'. It is the role of conscience to 'determine' these actions.[66] The conscience acts like an internal replacement of the Catholic priest: it is, Perkins wrote, 'a thing placed of God in the middest betweene him and man, as an arbitratour to giue sentence and to pronounce either with man or against man vnto God'.[67] It is the place where the Christian can find assurance of salvation. It enacts regeneration.

The tripartite structure of the human understanding is mirrored in the tripartite structure of the practical syllogism (or reflex act[68]) which is used to illustrate faith:

> *Every one that beleeues is the childe of God:*
> *But I doe beleeue:*
> *Therefore I am the child of God.*[69]

The first part is an act of mind – a general principle; the second part calls in memory – a personal and particular understanding; and the third part is a judgement which brings the first two together. Perkins links immortality – the existence of grace – with reason. The ability to syllogise represents the true exercise of the conscience, and the conscience itself can be seen as proof of God's grace in the Christian. The emphasis on education – whether Dorothy Leigh's literacy, or the learning of the catechism – is wholly tied up with faith and with salvation.

The assurance which comes with the true operation of conscience, the ability to reason, was needed by the Reformed Christian for two main reasons. Good works no longer counted for anything in the salvation of the individual: the believer could not work for his or her own salvation. But there was also and more problematically the belief in temporary faith; in a faith which even the reprobate might have, which looks to all intents and purposes like justifying faith but does not lead to salvation. Holland defines it as the ability 'to professe the Gospell but without sense of the power of it, or loue or liking to it'. If this were the case, it might be argued, then temporary faith would always be recognisable. But Holland continues: 'this faith may proceed yet to a great reioycing and to some heuenly fruite, and yet is it but false and temporarie.'[70] How to tell the difference between justifying faith and temporary faith was a problem which exercised Reformed thinkers.[71]

Calvin himself had warned of the dangers of the interrogation of status which would manifest itself in an examination of the 'hidden recesses of the divine wisdom'. He wrote, 'When once this thought has taken possession of any individual, it keeps him perpetually miserable, subjects him to dire torment, or throws him into a state of complete stupor.' But Calvin saw this as an inevitable product of man's fallen nature: 'it is right that the stupidity of the human mind should be punished with fearful destruction whenever it attempts to rise in its own strength to the height of divine wisdom.'[72] The search of the workings of God was futile and it was dangerous.

In his clearly titled *A Treatise Tending unto a Declaration Whether a Man be in the Estate of Damnation or in the Estate of Grace*, Perkins argued that '*there is such a similitude & affinity*' between the saved and the reprobate that '*it is the dutie of euery Christian to trie & examine himselfe, whether he be in the faith or not.*'[73] Perkins reiterated this way around Calvin's sense of the dangers of interrogating God in *The Foundation of the Christian Religion*. In answer to the question 'How may a man know that he is iustified before God[?]' Perkins wrote 'Hee neede not ascend into heauen to search the secret counsell of God: but rather descend into his owne heart to search whether *he be sanctified* or not.'[74] As Roy Porter has noted, 'Protestantism forced believers to go in for soul-searching', and the nature of the examination of the self was founded upon the operation of the conscience.[75] Knowing the nature of your own actions was central to understanding your own state: Perkins wrote 'in that you are grieued with a godly sorrow for your sins, it is a good token of the grace of God in you'.[76] The knowledge that you mourn your sins reiterates the sense of humanity as both sinful and saved, but also shows clearly the inseparability of reason and salvation as this knowledge could only come through the exercise of the conscience. Knowledge is both a gift from God and proof of God's gift: to know the self is to know God. But the conscience is not available to all; the animal does not possess one and the exclusivity of conscience is another form of assurance: the assurance of the separation from the beast.

However, such a simple distinction falls apart almost immediately. In *A Discourse of Conscience* Perkins wrote:

> Men commonly thinke that if they keepe themselues from periurie, blasphemie, murther, theft, whoredome, all is well with the[m]: but the truth is, that so long as they liue in ignorance, they want right and true directio[n] of conscience out of Gods word, and therfore their best actions are sinnes, euen their eating & drinking, their sleeping and waking, their buying and selling, their speech and silence, yea their praying and seruing of God. For they do these actions either of custome, or example, or necessitie, as beasts doe, & not of faith: because they know not Gods will touching things to be done or left vndone.[77]

The image of the beast reveals the importance of understanding in the establishment of the status of the Christian. Acting because of custom, without thinking, is unreasonable, and is not human. Perkins undermines the unthought sense of the human as always-already different from the animal and reveals that unthinkingness is not a sign of human-ness but of its opposite, animality. 'Men commonly thinke' like beasts 'because they know not Gods will'. But if knowing God's will is something the human must do for him- or herself – rather than relying on the always-alreadiness of faith – then the issues of predestination and double justification emerge in another way. The will of God must be learned, it is necessary to achieve human status; ignorance of God's will means a lack of humanity. Reformed thinking seems to pull in two directions: faith is God-given and unquestionable, but knowledge of that faith, which in many ways is faith itself, is human and not divine. Double justification gives the human a power which the belief in predestination seems to deny. Without regeneration the human is merely animal.

This is not an isolated use of the animal in Perkins' writing on the conscience. In *A Discourse of Conscience* the beast emerges on a number of occasions to figure everything that is the antithesis of the good Christian to the extent that 'daungerous sins' are likened to 'wild beasts [which] make hauocke of the soule.'[78] But the animal also exists in a literal and not figurative sense. At first Perkins makes the point about the absolute otherness of the animal very simply: where animals lack conscience all *'reasonable creatures'* (humans) have a conscience. Perkins warns against thinking 'that some men by nature haue conscience in them, some none at all' and proposes 'For as many men as there are, so many consciences there be: & euery particular man hath his owne particular conscience.'[79] But in this reiteration of the absolute distinction between all men and all animals can also be traced the perilous nature of the separation. Perkins goes on; 'Let Atheists barke against this as long as they will: they haue that in them that will conuince them of the truth of the Godhead, will they nill they, either in life or death.'[80] Atheists, the deniers of God, bark; they speak in the language of animals because in their atheism they reveal their lack of conscience. But the atheist's lack of knowledge is presented by Perkins as a choice – 'will they nill they'. Atheists 'haue that in them' to make them Christians, but they deny it. At some point there is a slippage. Where choice is denied in one instance – baptism – in another – atheism – choice is all, otherwise God has made humans into animals, a creation which would absolutely destroy the distinction between the species. It is as if the seed of faith which God has planted has never been watered; as if education, watering itself, is the thing which makes faith.

The image of the atheist barking is not only traced in Perkins' work. Thomas Beard tells the story of Lucian, who 'fell away' from his Christian faith and became an atheist. In this case the punishment truly fits the crime: 'This wretch, as hee barked out (like a foule mouthed dog) bitter taunts against the religion of Christ, seeking to rent it and abolish it: so he was himselfe in Gods vengeance torne to pieces and deuoured of dogs.'[81] Later in the text Beard speaks of atheists 'liuing in this world like bruit beasts, & like dogs and swine, wallowing in all sensuality'[82]. But, most famously,

Beard describes the life and death of the playwright Christopher
Marlowe:

> hee denied God and his Sonne Christ, and not only in word blas-
> phemed the trinitie, but also (as it is credibly reported) wrote
> bookes against it, affirming our sauiour to be but a deceiuer, and
> *Moses* to be but a coniurer and seducer of the people, and the
> holy Bible to be but vaine and idle stories, and all religion but a
> deuice of pollicie. But see what a hooke the Lord put in the nos-
> thrils of this barking dogge: It so fell out, that in London streets as
> he purposed to stab one whome hee ought a grudge vnto with
> his dagger, the other party perceiuing so auoided the stroke, that
> withall catching hold of his wrest, he stabbed his owne dagger
> into his owne head ...[83]

To say that God made Christopher Marlowe an atheist would be to
misread the tone of horror and fury in Beard's brief biography. The
reader is meant to be appalled at Marlowe's thoughts (which are
very much Marlowe's own, and not God's) and be relieved to see
God's judgement in action. Self-murder is an even more appropriate punishment than murder itself.

However, in *Christian Oeconomy* Perkins underlines the problems
the sense of atheism as self-willed proposes when he outlines a different interpretation. He argues that atheistic families are 'fitly com-
pared to an heard of swine which are alwaies feeding vpon the
maste with greedinesse, but neuer looke vp to the hand that beateth
it down, nor to the tree from whence it falleth.'[84] What is important
here is that atheism is presented not as a refusal of God, but as
ignorance of the workings of the world. It is the lack of understanding of the swine, rather than their willed departure from God,
which limits their potential for immortality. Atheism represents a
challenge to the Reformed sense of the absolute nature of the
distinction of the human from the unreasonable and conscience-less
animal. Perkins offers two versions: in one the atheist is a wilful
scorner of true faith, in the other God does not reveal himself.
The difference between these two bespeaks a confusion within
Reformed ideas about the status and potential of the human.

In *A Treatise Tending unto a Declaration* Perkins proposed that God
'hath made mee for his owne image, hauing a reasonable soule,
body, shape; where he might haue made me a Toad, a Serpent, a
Swine, deformed, franticke.'[85] It is because of the divine that Perkins

is saved. But in *Discourse on Conscience* Perkins also notes that 'Defect of reason or vnderstanding in crazed braines' would bring the wall of difference between the animal and the human crashing down.[86] Once again two slightly different ideas emerge: God's choice – 'he might haue made me a Toad' – and human incapacity – 'Defect of reason'. Atheist, toad and frantic emerge as different but equivalent terms, and all highlight the problem of asserting conscience as the feature which distinguishes human from beast and makes the difference between speaking and barking. There is the possibility of an act of God, but there is also the will of the human. These issues emerge again in the full embodiment of the barking human, the werewolf, and here we can trace some more difficulties in the Reformed notion of conscience.

IV

There are no cases of werewolves recorded in early modern England. Adam Douglas argues that the existence of real wolves and the existence of the mythical creature appear to go together, and that the extinction of the wolf on English soil by the end of the fifteenth century may well account for the lack of native werewolves.[87] What is also clear, however, from the frequency of references to werewolves in English writings, is the continued popularity of the idea of transformation: George Gifford writes that they occur 'In Germany and other countries', something which does not make them any less interesting.[88]

The narrative of Stubbe Peeter, reproduced in a pamphlet in 1590, (his real name was Peter Stumpf[89]) is one example of the popular transmission of the belief in lycanthropy. Stubbe Peeter, the werewolf of Cologne, killed women and children (and, cunningly, lambs 'as if he had beene a naturall Woolfe indeed'.)[90] He was caught when he was spotted removing the 'girdle' which caused him to be miraculously transformed from his wolfish shape into human form. At his trial in 1589 the judge noted that the girdle had mysteriously vanished, 'gone to the deuil from whence it came': Stubbe Peeter, like Faustus, is abandoned by Satan to his fate.[91]

What comes to the fore in this pamphlet is that there was no sense that Peeter at any time during his transformations ceased to be Stubbe Peeter. In fact, it was the continuation of his identity when in the form of the wolf which made him such a grotesque sinner. Had he been truly lost in the metamorphosis, had he ceased to be Peeter, then his innocence of the actions of the wolf could have been maintained. As it was, Peeter, 'fearing the torture', confessed to his crimes (he had killed thirteen children and two pregnant women as well as his own son), and was burnt at the stake.[92] But Peeter did not confess to himself; instead, he confessed to his judges, and judgement took place outside the individual. This was no true understanding of self, an act of conscience, but, paradoxically, an act of self-defence.

What this tale reveals is the sense in which the human remains a human even when undergoing metamorphosis: identity, in some fundamental way, is not altered. The conscience for William Perkins and other Reformed thinkers was an innate part of the individual, however, the story of Stubbe Peeter proposes that either such acts could be performed with the conscience in place, or that the conscience – the thing which separates human from animal – could be in some way abandoned so that Peeter could commit multiple murder, while the individual – Peeter again – remained intact. For Perkins both suggestions present problems. In the former there is a sense that the conscience, the God-given determining force, can determine incorrectly: that God could be, in a sense, immoral. The latter would also seem to be impossible, because there is no possibility of being human and having no conscience, the two are inseparable. Perkins proposes a possible way around this: the conscience, he argues, could be lost temporarily, 'onely in respect of the vse thereof, as reason is lost in the drunken man, and not otherwise'.[93] Peeter, intoxicated by the devil rather than the liquid served in those 'temples of the Devill', taverns, loses himself temporarily and, it would seem, deliberately.[94] But Perkins' explanation of the loss of conscience does not solve the problem of the status of the werewolf: if the conscience is where we prove our status as human and we can lose our consciences (albeit temporarily), then we can, by logical extension (that is, through the exercise of a practical syllogism no less), lose our status. The dividing line between the species vanishes, or at least becomes moveable. Peeter is at one moment human, at another animal. There is no stability in the difference.

The main debate about the werewolf in early modern England founded upon the nature of the transformation. Augustine's view, which would have been central to Reformed ideas, was summarised in John Bulwer's *Anthropometamorphosis* as 'a fascination'. 'The body of man cannot any way, by the Art or Power of Devils, be truly and really converted into the members and lineaments of a beast, but only the phantasticall appearance of a man.'[95] Neither the body nor the soul of the human subject were transformed, but a demonically inspired 'spectral double' went forth 'while the person lay asleep dreaming the things encountered by the phantom.'[96] These ideas, as Bulwer notes, 'are confirmed by *Aquinas*'.[97] As with the issue of the treatment of animals, there is some agreement between Thomist and Augustinian ideas here.

The argument about transformation re-emerges in the work of the sixteenth-century French judge Henri Boguet who writes, in agreement with Reformed ideas, that 'it is impossible for the body of a brute beast to contain a reasoning soul.' He also argues that it is impossible for the soul to be lost during the metamorphosis, asking the very obvious but important question, 'how is it possible for him to recover it, and for it to return into him when he resumes the shape of a man?'[98] The notion of the soul existing outside of the body, and yet the body still living, is inconceivable. But Boguet does not propose that werewolves do not exist; rather, he agrees with Augustine and Aquinas.

For Boguet's contemporary Jean Bodin, however, the transformation of the human into an animal in lycanthropia was a real one. The body actually physically changed. This could take place, Bodin argued, because 'the real essence of a human being ... was not physical form, but the rational faculty'.[99] For Perkins there would seem to be no argument with such an idea. Conscience – humanness – is a mental and not a physical thing. However, this interpretation of lycanthropia enraged Reginald Scot. In *The discouerie of witchcraft* he argued that

> whosoeuer beleeueth, that anie creature can be made or changed into better or worsse, or transformed into anie other shape, or into any other similitude, by anie other than by God himselfe the creator of all things, without all doubt is an infidell, and worsse than a pagan: because they attribute that to a creature, which onelie belongeth to God the creator of all things.[100]

Scot's dismissal of actual transformation is not merely a rejection of the supernatural in line with the rest of his work, it is also a defence of the human. 'These examples and reasons might put vs in doubt, that euerie asse, woolfe, or cat that we see, were a man, a woman, or a child. I maruell that no man vseth this distinction in the definition of a man.'[101] Bodin's interpretation of lycanthropia, which so closely, if accidentally, echoes Perkins' notion of human-ness, destroys the human. The separation from the beast on which being human is premised is impossible if boundaries can constantly shift. Scot's final insult to Bodin is a scorching one: 'I thinke it is an easier matter, to turne *Bodins* reason into the reason of an asse, than his bodie into the shape of a sheepe: which he saith is an easie matter'.[102] The mind is the frail part of humanity, the body – that which Aquinas said was all that an animal was made up of – is what Reginald Scot offers as the stronghold. Humans, in his interpretation, become wholly corporeal beings.

By the end of the sixteenth century, however, lycanthropia was changing. In August 1598 the case of Jacques Roulet in France revealed a new understanding. Roulet, a werewolf, had his death sentence commuted to a two-year sentence in an insane asylum, where it was ordered he should be given religious instruction.[103] Similar cases, where the guilty party was treated as mad and not bad, occur after this date. The creature, it would seem, was no longer the product of a fleshy, demonic transformation, but of an unstable mind. Physicians proposed that the terrifying creatures were, in reality, melancholics suffering from lycanthropia. In 1590 Henry Holland proposed that 'I denie not, but witches may haue also sundrie such Sathanicall delusions in many, which abounde in melancholy, but no reall transformations indeede.'[104] James VI and I presented 'Men-woolfes' as humans suffering from a 'naturall super-aboundance of Melancholie' which makes them 'become beastes by a strong apprehension'.[105] And John Deacon and John Walker, colleagues of Samuel Harsnett, wrote that lycanthropy was linked with 'disordered melancholy, ... mania, ... the epilepsy, ... lunacy'.[106] In John Webster's *Duchess of Malfi* the lycanthropic Duke Ferdinand speaks and reasons while in the throes of the illness. This could be merely dramatic technique, but could be further recognition of the apparent normality of the victim of the illness. Apart from this the symptoms Ferdinand displays are totally conventional.[107] In 1621 Robert Burton wrote of lycanthropia simply 'I should rather referre it to *Madnesse*, as most doe.'[108] The loss of mind and the descent into animality seem to underline Perkins' ideas, but they also undercut his sense of a clear status which can be termed human. If the conscience is the site of faith and the conscience can be lost then it is no longer possible to assert that human is distinct from beast. Reformed ideas, in their clear designation of human-ness, produce instead animality.

But the atheist and the werewolf are not the only figures to complicate the notion of the human in early modern religious writings. Many of the issues raised in Reformed ideas – baptism, education, conscience – emerge once again in problematic fashion in the figure of the wildman. This time it is not only the weaker members of human society – the female, the poor, the aged – who are shown to be dangerously close to the beast, the very men who write the words on which civilisations are produced are shown to be possessors of a fragile humanity.[109] The beast moves ever closer.

V

Records of wild children date from 1344 and the discovery of the wolf child of Hesse. Although the veracity of the narratives of their lives is dubious there is a sense of a real debate going on in the writings dealing with wild children, something which Lucien Malson argues should stop us from dismissing the tales out of hand, but should also remind us of the problematic nature of the line between myth and reality: '[t]otal acceptance would be just as wrong as total rejection'.[110] One of the most notable things to emerge from a study of documents relating to wild children is the number of repeated themes. These are themes which re-emerge in the tales of wildmen: fact feeds into fiction, fiction feeds into fact.

In his report of the case of John of Liège, which comes in a section dealing with the sense of smell, Sir Kenelm Digby, one of the first English Cartesians, raises a number of issues which resound throughout the present discussion, and for this reason it is worth quoting at length.[111] John was lost in the 'forest of Ardenne' as a 'litle boy' when hiding from enemy soldiers. After he was given up as lost by the community 'he lived many yeares in the woods, feeding upon rootes, and wild fruites, and maste [acorns].' Digby goes on,

> after he had beene some time in this wild habitation, he could by the smell judge of the tast of any thing that was to be eaten: and ... he could att a great distance wind by his nose, where wholesome fruites or rootes did grow. In this state he continued ...

untill in a very sharpe winter, that many beastes of the forest perished for want of foode, necessity brought him so much
confidence, that leaving the wild places of the forest, remote from
all peoples dwellinges, he would in the evenings steale among
cattle that were tethered, especially the swine, and among them,
gleane that which served to sustaine wretchedly his miserable
life. He could not do this so cunningly, but that returning often to
it, he was upon a time espyed: and they who saw a beast of so
strange a shape (for such they tooke him to be; he being naked
and all overgrowne with haire) beleeving him to be a satyr or
some such prodigious creature as the recounters of rare accidents
tell us of; layed wayte to apprehend him. But he that winded
them as farre off, as any beast could do, still avoyded them, till att
the length, they layed snares for him, and tooke the wind so ad-
vantagiously of him, that they caught him: and then, soone perceived he was a man; though he had quite forgotten the use of all
language: but by his gestures and cryes, he expressed the greatest
affrightedness that might be ... .

This man within a litle while after he came to good keeping
and full feeding, quite lost that acutenesse of smelling which for-
merly governed him in his taste; and grew to be in that particular
as other ordinary men were. But att his first living with other
people, a woman that had compassion of him to see a man so
neare like a beast; and that [had] no language to call for what he
wished or needed to have; tooke particular care of him, and was
always very sollicitous to see him furnished with what he
wanted: which made him so apply himselfe unto her in all his oc-
curents, that whensoever he stood in neede of ought, if she were
out of the way, and were gone abroad into the fieldes, or to any
other village neere by, he would hunt her out presently by his
sent, in such sort as with us those dogges use to do which are
taught to draw dry foote.[112]

There are a number of issues raised by this passage. First of all
John's transformation, like the transformation of the werewolf, is
temporary. Human status, although losable, can be reclaimed. But
there is also a sense in Digby's work, which fits his Cartesian beliefs,
that being human is revealed through the ability to speak. Digby
underlines John's true reclamation at the end of his narrative when
he writes that 'I imagine he is yet alive to tell a better story of

himselfe then I have done': John's return is figured through language, through his ability to tell his own story.[113] Descartes proposed that

> it is particularly noteworthy that there are no men so dull-witted and stupid, not even imbeciles, who are incapable of arranging together different words, and of composing discourse by which to make their thoughts understood; and that, on the contrary, there is no other animal, however perfect and whatever excellent dispositions it has at birth, which can do the same.[114]

From Digby's arrangement of the material about the wild boy it would seem that he follows this clear division between the species. Once John had regained language his acute sense of smell, which linked him dangerously with the animal, ceased to exist. John was either human (speaking) or animal (smelling), only for a limited time was he in some way both, and then the power of smell decreased as the power of speech increased. John could never be both true human and true animal; the two seem to be exclusive categories. But this, of course, is not true. If a boy can *become* a beast then exclusivity has vanished.

The Cartesian distinction of speaking and not speaking can be traced in earlier Reformed ideas about conscience, reason and barking. The atheist's loss of human language prefigures the division Descartes noted. Digby's allegiance to Descartes and his known recusancy do not totally separate him from Reformed ideas. Just as links can be made between Augustine and Aquinas in their ideas about werewolves, so links can be made between Perkins and Descartes.

But the tale of John of Liège offers another problem for religious ideas. Even the short time in which he was both human and animal, speaking and smelling, reveals, again, that the clear distinction between human and animal in Reformed belief could not work. The wild child regains human-ness by stages. In other renditions of cases of wild children – notably Bernard Connor's later work – the gradual nature of the return is a significant feature of the narratives.[115] But the lack of an exclusive human status to be traced here re-emerges in narratives of wildmen. Where the wild boy, or wolf child, might be based, ultimately, on a real case, the wildman was a fictional creation of a very different kind, and the image of the wildman is an important one in any discussion of the nature of the division between the human and the animal. The wildman was figured in a number of different ways – as ape, demon, savage, Irish native, New World native – to represent a border figure which made concrete the fear of descent into the animal.[116] Existing on the margins of society, the wildman is a reminder of the closeness of the beast.[117] But with this fear of descent there is also a sense of the possibility of return. Just as no-one ever questions the need to return the wild child to society, never questions that the quality of life for the child will be better in the so-called civilised world of human affairs, so the story of *Valentine and Orson* offers a sense of the dangerous and destructive possibility of descent to the status of the beast and an image of human separation from the animal.

The dissemination of *Valentine and Orson* is important here: the plot originated as a poem in the fourteenth century, and was rewritten as a French romance, printed in 1475 and 1489, eventually arriving in English translation in Wynkyn de Worde's edition of *c.*1505. The text then underwent three further editions in English; William Copland's two editions of *c.*1548 and 1565; and finally a rewritten edition printed by Thomas Purfoot in 1637.[110] *Valentine and Orson* is a narrative which has travelled across national and religious boundaries, and the changes to the narrative – I will be concentrating particularly on the 1565 and 1637 editions – are worth examining in the context of this discussion of Reformed ideas.[119]

The text tells the story of the twin sons of Alexander, the Emperor of Greece, and Bellysant, sister of King Pepin of France. Having been falsely accused of adultery Bellysant flees Greece and attempts to return to her brother's court, but she goes into labour with the twins in a forest in France. One of the babies, Orson, is stolen by a bear and the other, Valentine, is discovered, apparently abandoned, by King Pepin who takes him back to the court with him. Valentine is raised as a true knight while Orson becomes a violent wildman. On these basic points the two texts are the same, but what has changed are the religious implications of the narrative. Not only are references to learning about the Virgin, making the sign of the cross, visiting the Pope to make confession and other such obvious throwbacks to the 1565 text's Catholic origins removed, in more subtle ways the 1637 edition represents a rewrite of the earlier text for Reformed readers.[120] For our purposes two rewritten moments are particularly significant: the civilising of Orson, and the return of Orson to the forest at the end of the text.

When Valentine enters the forest to challenge Orson he does not realise that he is confronting his brother. It seems appropriate that Orson, whose viciousness debars all from entering the forest, and who has won every previous challenge offered to him, should lose to his twin. Importantly, however, he only loses through the power of persuasion. In 1565 this moment of verbal seduction comes when Valentine says:

> Alas wilde man, wherfore doest thou not yelde the unto me thou lyuest here in this wodde lyke a beaste, and hathe no knowledge of God, nor of his blyssed mother saynt Mary, nor of his holy fayth, for the whiche thy soule is in great daunger. Come on thy wai with me a[n]d then shalt thou do wyseli. I shall make the be baptized, and shall teache the, the holy fayth.[121]

The soul of Orson is at stake. The fight between the brothers is a fight for eternal life. Without learning, without knowledge, Orson will be damned; in this Pre-Reformation text Orson can work to achieve his own salvation. We are back to the Pelagian position.

The seduction of Orson in the 1637 edition is significantly different: 'Wild-man, wherefore dost not thou yeeld thy selfe to me? Heere thou livest like a Beast, having no knowledge of humane society. Come thy way with me, and I shall make thee know both thy selfe and others.'[122] There is no mention of God. The implication is that God is always-already present in Orson: that he is in receipt of grace. What has been dangerously absent from Orson's life is society; what must be learned is self-knowledge. Orson must interrogate himself to become a real man rather than remain an unselfconscious wildman. This would seem to fit in with Perkins' ideas about the conscience as the priest within, but baptism, which in the 1565 text is at the heart of the conversion of Orson from wild to civil, is not mentioned in the 1637 text until later, and when it does occur emphasis is placed not on scripture, but on the issue of naming (which Perkins says is the first role of the parent), and on the earthly role of the Godparent, Valentine. The emphasis is remarkably non-religious.

> The Baptisme being solemnized, the king sate him downe to Dinner ... *Orson* seeing the meate tooke as much as hee could graspe in his hand, and devoured it. Having eaten that he espyed one of the attendants that brought a Peacocke to the Table, who

comming neare vnto him, snatched away the Peacocke, and sat
him down and devoured it most greedily. *Valentine* seeing his behaviour, made signes vnto him that he did not well, wherat *Orson*
seemed ashamed ...[123]

As a Godparent, Valentine does not teach Orson to read the Bible,
or learn the Lord's Prayer, he teaches him manners, and it is
through Orson's realisation of his ignorance of social graces that we
glimpse the new workings of his conscience: '*Orson* seemed
ashamed'. It is not the presence of God which is lost, that is always-
already a part of Orson, what is lost are the skills of society and
what is needed is a polite education.

This sense of the this-worldly education of Orson is emphasised
later on in the text when Orson, who remains mute, meets Lady
Fezon. They fall in love in archetypal (if silent) romance fashion, but
'*Orson* made signes vnto the Lady, that he would never lie with her
till he had gotten use of his tongue'.[124] While Orson has been baptised and has learned the rules of society and can live within them
he is still an outsider: he makes signs, he does not speak. He cannot
marry – something which is no longer a sacrament, but is rather the
sign of integration into a human relationship – because he cannot
communicate with the voice. Orson needs language to be fully
civilised, and it is of course Valentine, his Godfather, who gives him
language: 'calling *Orson* vnto him, he cut the thred from vnder his
tongue that hindred his speech; which being done, he spake
presently, where he related the story of his life led in the Forrest'.[125]
'In early modern times,' Lloyd De Mause writes, 'the string under-
neath the new-born's tongue was usually cut, often with the
midwife's fingernail, a sort of miniature circumcision.'[126] Just as circumcision represents full entry into the Jewish community for the
male infant, so this snip of skin represents full entry into the human
community for Orson within the rewritten text.

After the snip Orson can explain himself, can tell the story of his
life, just as John of Liège can tell the tale of his return to civilisation
in Digby's report. Telling your own story is symbolic of being in
society. In fact, Orson's gaining of language, while it displaces
baptism as his point of entry into society, is also a model of the
Reformed understanding of christening. Before language there is
not a total exclusion from society and from communication, as his
sign language shows, there is just an incomplete inclusion which is
completed with the ability to speak. In Reformed ideas the child or

convertee is always-already in receipt of grace but is fully integrated into the Christian church through baptism.

The dangers of the definition of human-ness in Reformed ideas are made clear in *Valentine and Orson*. In the 1637 text speech has displaced the conscience as the marker of human status because the assertion of the conscience as the site of difference between human and animal is no longer possible: the atheist and the werewolf have proved this. Instead, Orson gains human status through the snip. The existence of faith is not the issue in the 1637 text, but instead the regenerative act of Reformed theology – education – becomes the thing itself. The discussion of the conscience is silenced.

But this sense of the returning human, the reclaimed member of society who is able to distance themselves from their past by narrating it, is undermined at the end of *Valentine and Orson*. After numerous adventures and Valentine's death, Orson returns to the forest, his original home. In the 1565 edition the return to the forest follows Orson's vision of heaven and of a rather Boschian hell – a place where, significantly, he sees people 'hanged by their tongues'. Hell silences, is punishment perhaps for incontinent speech. The vision leads Orson to take up the contemplative life; he 'lyued holyly, & after his deathe god shewed for him many myracles and was a saynt canonyzed.'[127]

Again, however, the 1637 edition makes some important changes. Orson has a vision, but this time it is beyond language, 'more glorious then his tongue could vtter' and it literally takes him out of human society.

> After this vision he awaked, and being astonied thereat, with teares he came to the Greene knight, and sayd: Sir, I see the vncertainty of the world, for which, I am willing to resigne my estate & Children vnto you: see them well educated, that they may by you be made fit to gouerne such a people, for I will spend the rest of my daies in solitary contemplation.[128]

Orson abandons his children and his worldly responsibilties: in baptismal fashion he hands over the infants to the substitute parent. But, most importantly and paradoxically, Orson abandons society, and, apparently, his status as a human.

But this is not what the text asks us to believe. In the 1565 edition Orson's return to the forest is the fulfilment of an ideal: he becomes a hermit and ultimately a saint. In the 1637 edition a similar evolution takes place. Orson's faith – his possession of a conscience – is never in doubt in this text, what is in doubt is his knowledge and it is a fulfilment of an ideal of human knowledge which Orson aspires to at the end of the reformed text. This new ideal could be termed Mirandolan rather than Reformed. In *On The Dignity of Man* Pico della Mirandola wrote:

> if you see a man given over to his belly and crawling upon the ground, it is a bush not a man that you see. If you see anyone blinded by the illusions of his empty and Calypso-like imagination siezed by the desire of scratching, and delivered over to the senses, it is a brute not a man that you see. If you come upon a philosopher winnowing out all things by right reason, he is a heavenly not an earthly animal. If you come upon a pure contemplator, ignorant of the body, banished to the innermost places of the mind, he is not an earthly, not a heavenly animal; he more superbly is a divinity clothed with human flesh.[129]

Pico offers the results of various choices: these range from the bush to the pure contemplator. The true fulfilment of man's dignity in Pico's 'manifesto of Renaissance Humanism' is the man who turns away from the world – the flesh – and contemplates. This is what Orson does in the 1637 text.[130]

The change from the Reformed to the Mirandolan view at the end of the text reveals some of the problems which this chapter has traced. Pico's man is the Reformed vision of the prelapsarian Adam. For Reformed thinkers the perfection of the original man which Pico proposes is impossible to achieve after the Fall, is lost to humans forever. The status of the divinity, the ability to approach the Godhead, is not available to the human acting alone within Reformed ideas. Instead what the Calvinist thinkers of England offer is a man of worm-like stature and ability who can only achieve salvation through the will of God.

The figure of the wildman, like the atheist and the werewolf before him, is a massive problem within Reformed notions of human-ness. Where Perkins ends by offering two very different interpretations of the atheist, and Scot reverses Perkins' designation of the conscience as the site of human-ness, the reformer of *Valentine and Orson* takes these difficulties to their logical extreme. There is no point in offering the conscience as the defining feature because it is so fragile. Some humans remain human without the use of conscience, and the wildman can always-already be saved. What differentiates the species is something very different. Speaking and understanding become the signifiers of humanity: Orson has learnt to speak, has entered human society, and his retreat to the forest at the end of the text represents the true and absolute expression of his human status.

It is to eloquence rather than baptism that the reformer of the tale of the wildman turns, not because there is no wish to fulfil the true reformation of the text, but because the truth which the Reformation has to offer is too painful, too dangerous to the species, the reader. To say that an animal lacks conscience is no longer meaningful. The logic of Reformed ideas reveal the impossibility of a stable status which can be termed human. Instead the idea of conscience, an invisible status, is replaced by the more determinable ability of speech. It is in eloquence that the human can be found.

3

Judging Like a Malt-Horse: The Humanist Interpretation of Humanity

'Speak that I may see you': Socrates' dictum, cited by both Erasmus in the early sixteenth century and Jonson in the early seventeenth, comes to life at the end of *Valentine and Orson*.[1] Orson becomes visible through his ability to communicate, and his contemplative state at the end of the text reveals a new notion of the species, one which can be termed humanist rather than Reformed. Speech and identity are inextricably linked: just as Orson needed to gain his voice to truly enter the human community, so humanists proposed that spoken communication was a signifier of humanity. This emphasis on speech would seem to offer a solution to the danger-ous frailty of human status offered by the Reformed emphasis on conscience. But at the same time as eloquence was emerging as the site of human-ness the question of interpretation was also an issue. Eloquence was only a signifier of the human if it could be under-stood; in fact, eloquence, like beauty, is in the eye of the beholder. The human can be as eloquent as is humanly possible, but if his eloquence is not understood, if he is not interpreted aright, then his eloquence counts for nothing, and as such interpretation becomes the skill which defines the human.

This chapter looks at the possibilities and problems in the differ-entiation of the human from the animal within humanist writings, but it begins with the Reformed vision of the body and the regener-ative activity of clothing. The emphasis on physicality within Reformed ideas of double justification is shown as a development of the importance of the conscience which was traced in the previous chapter and it offers a place where English humanism – which is concerned with the mind – distinguishes itself from its religious context. But this distinction also exists in another area. The repeti-tion of catechisms is mirrored in the humanist learning by rote of the classics, but in humanism there is a need to read another language and to allow for this a new form of learning is introduced which brings into play the animal.

William Rankins wrote, 'A shame it is ... to humanitie, that brutish beasts, wanting reason, should instruct men'.[2] This shameful activity is exactly what takes place in humanism: the animal – the epitome of ignorance – is used to teach the human through fables. The beast fable is not merely a literary convention, it actually enacts the aim of humanism itself. To look beneath the surface of the fable, to read the moral and not the animal, is where the human can be found. As an illustration of this I examine Philisides' beast fable in Sir Philip Sidney's *Old Arcadia*. In this fable the animal not only disguises the humanist moral; it also reveals a notion of interpretation which defines human-ness itself. To misread a fable is to be an animal.

For Ben Jonson, however, the practice of being human within humanism is not so much in the written as the spoken word: speaking is the site of the human. This is a distinction which Jonson recognises and responds to, and his sense of the difference between the theory and practice of humanism emerges in *Volpone*. But Jonson extends Sidney's theory into practice to reveal that the theory can only stand as a theory. Interpretation is revealed as impossible in the face of true eloquence. Reformed and humanist ideas assert two very different notions of the human – one based on the conscience, and the other on interpretation – but both come to the same conclusion: there is no always-already human, there is only human-ness, a quality which must be learned, and can be lost.

The link between Reformed and humanist ideas is made clear in the work of William Prynne. The body, the corporeality which humanism attempts to ignore, returns to the centre of the debate, and being human in the theatre is revealed as impossible in a way which extends both Jonson's and the earlier anti-theatricalists' logic. By the end of the chapter neither the mind nor the body are sources of human-ness. The animal, in fables and on stage, reveals difference to be a fiction. Just as the reformed text of *Valentine and Orson* reveals the fragility of the Reformed notion of the human – the possessor of a conscience – so the Reformed anti-theatrics of Munday, Stubbes and Prynne, and the humanist theatrics of Jonson reveal Sidney's understanding of the human as an eloquent and interpreting being to be purely theoretical. This chapter begins, however, with the thing which humanism will attempt to abandon: the body.

I

In I Corinthians, 3: 16–17 is written 'Know ye not that ye are the temple of God, and that the spirit of God dwelleth in you? / If any man shall defile the temple of God, him shall God destroy; for the temple of God is holy, which temple ye are.' Paul's exaltation of the human frame is repeated in numerous Reformed works. Thomas Morton wrote, 'As the fall of man defaced the Image of God in man; so the grace of God doth wholly restore the same, both in happinesse and also in holinesse.'[3] The body is a temple; it has been damaged by the Fall but it can be restored. However, the inverse is also true: to sully the body is to reveal an absence of grace, is to display reprobation. In *An Exposition of the Symbole* William Perkins wrote

> mans body by creation, was made a temple framed by Gods own hands for himselfe to dwell in; therefore our duty is to keepe our bodies pure and cleane, and not to suffer them to be instruments, whereby to practise the sinne of the heart .... For the more filthie a mans body is, the more fit it is to be a dwelling place for sinne and Satan.[4]

This acknowledgement of the sacredness of the body is a response to the invisibility of grace. To dress the body correctly is to enact regeneration, whereas as to possess a 'filthie' body is to reveal visibly one's status as reprobate. Clothing and salvation are linked within Reformed theology.

This is the absolute antithesis of Pico della Mirandola's representation of man. The highest human ambition, he argues, is sought in the mind: it is the 'divinity', the pure contemplator, who is Pico's ideal human – he has abandoned the flesh.[5] For Reformed thinkers, however, the body remained central. In the postlapsarian world it was not to be ignored but maintained with human effort: clothing had to be provided. After eating the forbidden fruit Adam and Eve immediately covered themselves with fig leaves, 'Then the Lorde pitiying their miserie, and loathyng their deformitie, gaue the[m] peltes, and felles of beastes, to make the[m] garmentes withall, to the end that their shamefull partes might lesse appeare'.[6] The Fall brings with it feelings of shame and – as Roger Crab was to note in the 1657 – an indebtedness to animals: 'they were beholding to other breathing creatures to uphold their bodies'.[7] Animals once again emerge as the status of humans is questioned. Where the conscience is the primary site of grace, dress becomes evidence of the operation of conscience. There are beasts in the conscience and there are, by extension, animals in clothing.

Adam Hill proposed that apparel was needed just to cover the 'unseemly parts' and to defend against the elements, but for Phillip Stubbes clothing was more than this: it was a 'signe distinctiue', setting apart the classes and the sexes.[8] In both of these readings clothing remained indifferent – *adiaphora*: it did not have a role above earthly interests. Robert Crowley said the same of priestly vestments, but added 'when the use of them will destroy or not edifie, then ceasse they to be so indifferent'.[9] Stubbes followed this line and fashion – 'exquisitenesse of Apparell' – emerged as the worst form of pride humans could exercise: 'as an exemplarie of evill, [it] induceth the whole manne to wickednesse and sinne.' Fashion was the antithesis of the Reformed idea of double justification: it was, in fact, evidence of both damnation and reprobation. Clothes were a product of the Fall, and fashion was evidence of the fallen and reprobate nature of the human.

But the issue of regeneration raises the problem of humans making themselves reprobate. There is a sense of choice – of wearing or not wearing – in the discussions of the regenerative power of clothing. Just as atheism is both an act of God and a position chosen by the human in Perkins' work, so in *The Anatomie of Abuses* fashionable apparel was an advertisement of the human's reprobation (God's work) and a temptation to others (human's). Stubbes wrote 'This sinne of excesse in Apparell, remaineth as an example of euill before our eyes, and is prouokatiue to sinne, as Experience daiely sheweth.'[10] Even within Reformed ideas there lurk the echoes of Pelagianism.

But ornate clothing enacted reprobation in a specific way. Fashion was quite simply, Stubbes argued, 'Sodomitrie', the most unnatural use of the human form. This was not the only form of sodomitry which he saw in late sixteenth-century society. Stubbes is most famous for his attack on cross-dressing on the stage, and it is in his diatribe against dramatic performance that the notion emerges once again.

Then these godly Pageantes beyng doen, euery mate sorts to his mate, euery one brynges an other homewarde of their waie very freendly, and in their secrete conclaues (couertly) thei plaie the

Sodomits, or worse. And these be the fruites of Plaies and Enterludes, for the most part.[11]

The gestures which are depicted in the theatre, like the very visible nature of fashion, are so powerful that they can destroy marriage by leading the spectator to sodomitical relations with a member of the same sex. But this is not the final degradation which Stubbes recognises; there is also a 'worse', a crime which can and will be committed following attendance at the theatre. In sexual terms in this period worse than sodomy can only mean one thing: bestiality. John Rainoldes repeated Stubbes' interpretation when he called the actor 'a *dogge*' and argued that plays were 'occasions whereby men are transformed into dogges'.[12] Bestiality is a crime which results not merely in the destruction of a human institution – marriage – but in the destruction of the human itself.

Fashionable clothing and the theatre find a link through the image of the animal. Both are bestialising. But the theatre is also important here in another way. If eloquence and interpretation are signifiers of human-ness, as humanism proposes that they are, then the theatre should have a role to play in creating the human. This is not the case, and the differences between Reformed and humanist ideas are usefully illustrated by Anthony Munday. In *A Second and Third Blast of Retrait from Plaies and Theaters* Munday wrote

There co[m]meth much euil in at the eares, but more at the eies, by these two open windowes death breaketh into the soule. Nothing entereth in more effectualie into the memorie, than that which commeth by seeing, things heard do lightly pass awaie ...[13]

Memory is, William Perkins argued, central to the operation of the Reformed understanding, and sight and hearing are linked to this faculty by Munday.[14] But for Munday seeing in the theatre is dangerous, it corrupts the memory, and listening is merely ephemeral. 'Speak that I may see you' fails miserably in Reformed antitheatrical ideas. Hearing, interpreting, is not a positive term for Munday it is merely a weak possibility which is always overridden by the visual. From the Reformed perspective the theatre is always-already dangerous, and humans are always-already fallen. All the Christian can do, in good Reformed fashion, is accept this imperfection and in their acceptance understand the just nature of God. For Munday the potential for achieving human status through interpretation is gone with the loss of Eden: the human is always-already lost.

Munday's other career as a professional playwright places him in a paradoxical position as an anti-theatricalist, and is something which links him to Ben Jonson.[15] It is, however, important also to trace a line from Munday to Stubbes and Perkins in the emphasis which he places on the importance of the visual. For Stubbes, writing three years after Munday, fashion is most dangerous to the immortal soul because it is most visible (and therefore tempting) to the fallen eye. For Perkins, the body must be kept clean because it is the temple of God, and filth is evidence of reprobation. Despite the emphasis upon the invisible – the conscience – the visible plays an important role in Reformed thought. In fact the impossibility of proving human status through the invisible, and the dangers of attempting to penetrate the word of God, create the context in which regenerative activities, such as correctness in apparel, become important.[16]

Again, however, a paradox arises. Where the visible is used to replace the failings of the invisible in the regenerative activities of dress the visible itself becomes highly dangerous. It is a throwback to the old religion, to popery. Robert Crowley wrote 'We are therefore that people must serve God in spirit and truth and not in figures and shadowes'.[17] But the importance of the visual is an inevitable result of the Fall: our eyes were opened with the eating of the forbidden fruit (Genesis 3: 5) and the first thing that was seen was nakedness (Genesis 3: 7). This link between seeing and the body, between knowledge and flesh, pervades early modern Reformed thought and is a link which is left behind in Mirandolan humanism. In Pico's work the abandonment of the flesh allows the human to reach up to the status of the divine: perfectability is not physical, it is mental. This disjunction is important: where in Reformed thought the human body is made in God's image and is a way in which we are distinguished, as Perkins argues, from toads, in humanism the physical body counts for nothing.[18] The distinction from the animal must be sought in another place, in the operation of the mind.

II

The appearance of the Bible in English in 1526 was perhaps the single most significant development of the sixteenth century: scripture became available in the language of the lay believer. In humanism, however, emphasis was placed upon a return to classical sources, on the reading of Latin and Greek texts in the original, and once a language other than English was introduced the process of education became very different.[19] In his translation of a collection of fables '*drawne out of the ancient writers*' Thomas North writes that 'Wordes teach but those that vnderstand / the language that they heare'.[20] The truths contained in written material are available only to those who know its language. As an illustration of his point North's collection contains a tale of 'the simple ignorant man, desirous to seeme learned' who learns 'by rote' some 'goodly sentences' which he repeats, 'not vnderstanding the signification of the words'. The man is scorned by the scholars who hear him and the moral is clear: '*Every man therefore must indeuour himselfe to vnderstand that he readeth*'.[21]

The importance of true understanding has slightly different implications in *The Second Tome of Homilies*. Here is written 'For Owles and Popiniayes, and Rauens, and pyes, and other suche lyke byrdes; are taught by men to prate they know not what. But to sing with vnderstanding, is geuen by Gods holy wyll to the nature of man.'[22] The birds can make the noises, but cannot understand the meaning. Understanding – presented here as a harmonious activity – is available only from God. This works within the Reformed notion of repetition, but in humanism the sense of the ability to 'sing with vnderstanding' becomes limited in its application. A new language – Latin, Greek or Hebrew – needs a new learning. The illiterate – 'unlettered', in Latin – is outside humanism, and if humanist pursuits fulfil human potential, make, in fact, the human truly human, then those who could not join in, who were not literate, would seem to be not human. The truly-human human of humanism is an elite category: one must know Latin to understand Latin.[23] Humanism seems to be making humans but it also reveals how frail and how created (as opposed to natural) being human is.

This sense of the unnaturalness of the human of humanism emerges with clarity in the educational practices which the ideas promoted. As in the case of catechisms in the Reformed church, reformed education emphasised repetition and the ability to memorise: Ovid's *Metamorphoses* was learnt by heart at the rate of 'twelve lines a week, five hundred lines a year, for two or more years'.[24] However, unlike the Reformed idea of repetition, the repetition of the Classics required not so much the grace of God as education. The printer's preface to John Clarke's *Phraseologia puerilis Anglo-Latina* advises that this collection of Latin phrases and English translations can be used by beginners who should 'learne all the phrases without booke, which done, let some one in a forme take the book and oppose his fellowes in the same forme, and thou shalt see how they will with great emulation and delight ambitiously strive to answer.'[25] This was not an innate, God-given ability, rather this was learned; the boys 'strive' to answer.

The notion of humanist learning as a labour can be traced throughout the early modern period. In 1581 the great Elizabethan school master, Richard Mulcaster, wrote that the 'Elementarie', as opposed to 'Grammarian', education was important because 'sufficiency in the child, before he passe thence, helpes the hole course of the after studie, and influencie slipping from thence to soone, makes a very weake sequele.'[26] Later in the text he reiterated this point with an emphasis on the importance of embedding learning deep in the mind of the child: 'For when the thinges, which be learned do cleaue so fast in memorie, as neither discontinuaunce can deface them, nor forgetfulnesse abolishe them: then is abilitie vpon ascent'.[27] In the same vein, in 1634, William Gouge wrote, 'The apprehension of children is fickle, and their memory weake: if they be but once, or seldome, or slightly instructed, that which is taught, will soone slip away, and do little or no good.'[28] In 1660 Charles Hoole reiterated this sense of the difficulty of learning:

> children in their tender years are generally like leaking vessels and no sooner do they receive any instructions in grammar but they forget them as quickly, till by frequent repetitions and examinations they be reveted into them and by assiduity of long practice brought to an habit, which cannot be bred in them under two or three years time ...[29]

Constant reiteration is necessary to instil knowledge.

This mode of repetition differs, as noted, from the Reformed sense of learning and a humanist prefiguration of Perkins' make-up of the understanding appears in Mulcaster's work. Where Perkins has mind, memory and conscience, Mulcaster proposes 'witte to take, memorie to keepe, discretion to discern' in the developing understanding of the scholar.[30] In place of Perkins' delineation of the Reformed conscience we have discernment, which for Perkins is something which must be obtained *before* conscience is in place: it is part of the 'preparation to good conscience'.[31] Discernment, it could be said, is human, conscience divine.

But in the learning of discernment, the quality which defines human-ness, we encounter once again the animal. Where humanism seems to assert an absolute distinction of the beast from the human it is, in fact, reliant on animals. At the heart of the humanist education was Aesop.[32] The beasts of the beast fable are the teachers of humanist ideals. Animals, it would appear, do humanism before the humans, and the human in humanism becomes strangely connected to the beast. Aesop's fables illustrate the two sides of the humanist endeavour: the operations of grammar and the importance of moral actions. In a 1624 edition of the fables John Brinsley defines their function:

> *First cause your scholar, by reading this translation, to tell you in euery Fable, what the matter of the fable is. Secondly, to what end and purpose it was inuented, what it is to teach, aad [sic] what wisedome hee can learne out of it. Thirdly, how to make a good report of the fable, both in English and Latine, especially in English. Fourthly and lastly, to make right vse of it, for all matters concerning Grammar, as for construing, parsing, making and prouing the Latine; and so for reading forth of English into Latine ...*[33]

It is ultimately contradictory that the animal should be so significant to an educational endeavour which asserts that the potential of the human is so much greater than that of the beast. This paradox, that human-ness – expressing correctly and acting morally – should be learned through animals, through what was to be left behind, was noted by Thomas North who traced the changing place of 'brute and dumb beasts' in the fable:

> In English now they teach vs wit.
>   In English now they say,
> Ye man, come learne of beastes to liue,
>   To rule and to obay,
> To guide you wisely in the world,
>   To know to shunne deceyt,
> To flee the crooked pathes of guile,
>   To keepe your doings strieght.[34]

The fable is, however, the genre of speaking beasts: the dumb creatures are not dumb, are neither silent nor stupid. More that this, the objectification of eloquent animals is a contradiction within humanist ideas which emphasise eloquence as evidence of human-ness. In the fables, however, it is just this objectification of animals which is, North proposes, their true strength: 'things, to men of sundrie speach, / Examples make appeare.'[35] The beasts, as R. W. Maslen has pointed out, speak 'a universal alternative language which can help to overcome the divisions between verbal languages.'[36] The animal undermines the Tower of Babel which has been erected in the humanist emphasis on reading classical texts in the original. It is an image of paradise regained. Through animals absolute human control can be figured. The human becomes, in John Moore's words, a 'petty god'.[37] The problems of the new learning – the emphasis on foreign tongues – are solved only through the animal, but the solution to the problem brings more problems. The thing which should be the antithesis of being human – the animal – becomes the means to achieve human status.

But, of course, the animals in beast fables are not animals at all, they are creations of the human imagination. The beast fable is a formal genre, an ancient and well established mode of expressing ideas. The real animal is clearly absent from Aesopic works, and a study of the fable would seem to have little to say about the perception of animals in early modern England. However, in the writings which support fables a sense of species difference can be traced. As in Reformed thought, while the animal is clearly in place – is clearly animal – it is humanity itself which makes problematic the notion of being human.

III

The literary and the interpretation of the literary were central to the humanist endeavour to educate. Poetry – a generic term for fiction at this time – was a place wherein interpretation could be performed, where the animal could be left behind. George Puttenham went as far as to argue that poetry was not a product of human civilisation, it had actually *created* it:

> The profession and vse of Poesie is most ancient from the beginning, and not as manie erroniously suppose, after, but before any ciuil society was among men. For it is written, that Poesie was th'originall cause and occasion of their first assemblies, when before the people remained in the woods and mountains,

vagarant and dispersed like the wild beasts, lawlesse and naked, or verie ill clad, and of all good and necessarie prouision for harbour or sustenance vtterly vnfurnished: so as they litle diffred for their maner of life, from the very brute beasts of the field.[38]

Where, as *Valentine and Orson* had proposed, wildernesses make wildmen (animals), poetry, Puttenham argues, makes civilisation and, by implication, scholars (humans).

It was not only Latin and Greek texts, or translations of ancient texts like Brinsley's *Esop*, which were to be interpreted by the scholar, however, vernacular writings, the works of contemporary poets, were also regarded as important. In his dedicatory epistle to Spenser's *The Shepheardes Calender* E. K. praises the poet for his restoration of the honour of the English language.[39] Similarly, in *A Defence of Poetry* Sir Philip Sidney singles out George Buchanan, the sixteenth-century Scottish humanist and poet, for praise alongside Spenser, Thomas Sackville and the Earl of Surrey.[40]

At this time works in the vernacular were also framed by didactic prefaces which declared both the nature of the poetry itself and the correct way of reading. For Thomas Nashe poetry was 'a more hidden and divine kind of philosophy, enwrapped in blind fables and dark stories'. Richard Mulcaster defined poetry as 'cover[ing] a truth with a fabulous veil'.[41] Poetry, like the fable, was a hidden form. The 'Prologue of the Second Booke' of *The Fables of Esope in English* states that 'fable is as much to say in poetry, as words in theology'.[42] Where Calvin warned against looking into the 'hidden recesses of the divine wisdom', poetry was in place to be trawled for meaning.[43] In 'The Prologue' to *The Morall Philosophie of Doni* the reader is told

> For alwayes the worke of these sage Fathers carieth two senses withall. The first, knowne & manifest. The second, hidden & secret. Of the first we sweetly enioy the taste: but of the second we receiue small knowledge, if we deeply ponder not the words.[44]

True meaning in poetry lies beneath the surface and has to be searched for, but ultimately this true meaning, the hidden object of interpretation, is available. But the elitism of humanism, the sense that it is limited to those with the right education, is emphasised in Henry Reynolds' *Mythomystes*. Writing of the '*Valve of Trve Poesy*', Reynolds echoes Calvin's warning about the interrogation of the

scriptures and underlines the sustained and problematic importance of Pico della Mirandola to English humanism. He writes

> But those secreter Mysteries, and abstrusities of most high
> diuinity, hidden and concealed vnder the barke, and rude couer
> of the words, to haue diuulged and layd these open to the vulgar,
> what had it been other than to giue holy things to dogs, and cast
> pearles among swine?[45]

This passage is in part a reference to Matthew 7: 6, but as a whole it
is an almost direct and unreferenced quotation from *On The Dignity
of Man*.[46] As a good humanist Reynolds invokes both the Bible and
the manifesto.

But the use of the animal image in the scriptural citation is important here as well. Giving holy things to dogs: the animal imagery
seems to be safe in its conventionality, it symbolises the unlettered
human. The dog is a metaphor not a true likeness. Such metaphors
are scattered throughout early modern writing about education and
interpretation. To offer just one example, John Moore, echoing Job
11: 12, asked 'What is an infant but a bruit beast in the shape of a
man? and what is a young youth, but (as it were) a wilde untamed
Asse-colt unbridled?'[47] Metaphorically speaking – '(as it were)' –
education was a bridling, a breaking in of the child.[48] The non-literal
nature of the use of animal imagery is made explicit in Thomas
Adams' sermon *Lycanthropy, Or, The Wolfe Worrying the Lambs* where
he writes that 'The nature of our duties is exemplified in this word,
Lambes. Not that there should be a *Metamorphosis* or transformation
of vs into that kinde of beasts, *literally*. But, *as Lambs*.'[49] In Adams'
text the lycanthropes are not actual werewolves, but

> mysticall *wolues*; rauenous beasts in the formes of men: hauing a
> greater similitude to *wolues* in the disposition of their mindes,
> then dissimilitude in the composition of their bodies. The wicked
> haue many resemblances to *wolues*.[50]

This is Reginald Scot's reading of lycanthropy rather than Jean
Bodin's. There is no bodily transformation here. In the same way,
animal imagery in the beast fable represents a kind of literary shorthand. There are, as in Moore's and Adams' Reformed uses of the
animal symbol, apparently no implications for the nature of
humans, there are only 'resemblances'. This, however, is not the

case for all; in the work of Sir Philip Sidney the metaphor breaks down.

Sidney's *Old Arcadia* is a Reformed humanist text: to read this text is to engage with the issues of late sixteenth-century English humanism in that the world it represents is not so much Mirandolan as Calvinist. In this text humans must accept their wretchedness and their place in God's universe: choice is not an option. *The Old Arcadia* is also a work, in the tradition of so much contemporary writing, which requires the reader to look beneath the surface. The good reader is required to make meaning. Such an emphasis on the sense of a 'correct' way of reading the text answers the problems which are thrown up by the actions of the two central characters. Pyrocles and Musidorus remain heroic in a text which includes Musidorus' only just averted rape of Pamela, and Pyrocles' unchivalric attempt to commit suicide because it is expected that the reader will notice these moral failings and apply their own judgements where the author might not do so. The failings in the heroic characters are not failings on Sidney's part, but are a part of the humanist endeavour: we are educated by thinking. Sidney's control of meaning is represented through the offer of alternatives, he is actually inviting the reader to discover the true meaning for themselves.[51]

The central narrative of *The Old Arcadia* deals with King Basilius' inability to interpret the oracle. Just as Calvin's providentialism proposes that the order of the universe is preordained, so the existence of the oracle means that certain events are inevitable. Basilius, however, thinks that he can avoid the prophesy. He believes, implicitly, that he can, like Pico della Mirandola's man, make his own world; that this is the power which the divine has given him. But Basilius' initial failure of interpretation, his misunderstanding of his place in this Reformed, and not Mirandolan world, opens the way for a number of other failures. Even the interpretation of true identity becomes problematic: Basilius thinks that Cleophila (the cross-dressed Pyrocles) is a real woman, while Queen Gynecia interprets correctly and recognises that Cleophila is a man. But in making the right interpretation Gynecia is not a true humanist. She fails to put her right interpretation to the correct use and encourages her husband's misinterpretation for her own ends. She is a good reader but a bad subject, and because humanism, as the prefatory materials which accompany beast fables propose, asks for the correct translation and a recognition of the moral, the right

interpretation without moral action is like reading without understanding. The right interpretation put to the correct moral use is
what is called for. A bad subject can never be a good humanist.

The end of the narrative of *The Old Arcadia* seems in many ways
unsatisfactory. Basilius, the bad reader and bad ruler is 'killed' by
his wife and subject, Gynecia, but comes back to life; Pyrocles and
Musidorus are rightly condemned for their actions and then acquit-
ted; but most importantly the truly just punishments which
Euarchus has meted out are overthrown. David Norbrook has
argued that Euarchus represents Sidney's fear of preciseness, of
Puritanism, I am arguing that Euarchus is both a good politician –
he knows the value of outward display – and an ideal judge.[52]
Euarchus enacts the humanist return to the source in his emphasis
on the letter of the law 'especially of Greece, and particularly of
Arcadia (wherein I must confess I am not unacquainted)'.[53] He also
refuses to be moved by the outward appearances of the Princes
because the true self lies within: 'thoughts and conceits', he says,
'are the very apparel of the mind'.[54] Where Reformed thinkers such
as Stubbes and Perkins regard dress as evidence of the status of the
Christian, and raise the ugly head of Pelagianism, Euarchus refuses
the dangers of such regenerative logic and proposes a more pure –
if such a term is appropriate – Reformed line: that thoughts, the
invisible, attest the visible. It is the inner self which, in this
Reformed humanist text, represents the true self. The good reader
who, like Euarchus, understands the distinction between the
surface and the underlying meaning will acknowledge the uncomfortable nature of the ending of *The Old Arcadia*; will question the
morality of the text; and will recognise that this questioning is part
of the endeavour of the work. We are forced to work to understand,
we are not offered meaning on a plate: in this *The Old Arcadia*
emerges as a truly humanist text.

Paul Oskar Kristeller reads humanism, Alister McGrath proposes,
as 'a cultural and educational movement, primarily concerned with
written and spoken eloquence, and only secondarily concerned
with matters of philosophy and politics.'[55] This separation of eloquence from politics is important in understanding the difference
between Sidney's *Old Arcadia* and Spenser's *Shepheardes Calender*. In
Spenser's text meaning is spelt out in E.K.'s glosses and the political
undercurrents become the only interpretation available to the
reader. For Sidney, however, meaning is to be found through effort.
Annabel Patterson's sense of the difference between *The Old Arcadia*

and *The Shepheardes Calender* offers a useful distinction here: she argues that Sidney uses the fable to criticise 'the *concept* of monarchy', while Spenser uses it to criticise 'the *practice*'.[56] Spenser's practical (glossed) application of the fable requires a different form of reading to Sidney's theoretical work. Spenser is forcing a meaning. The liberty of reader-response which is so much a part of Sidney's endeavour is denied in Spenser's text.[57] As an illustration, in *The Old Arcadia* there is a poem which takes humanism to its logical extreme. Philisides' beast fable, which appears between the Third and Fourth Books, presents interpretation as a matter of importance in the establishment of the human.[58] An incorrect reading makes the reader literally animal. Where Sidney is, according to Patterson, looking at the '*concept* of monarchy' in his work, he is also, I suggest, looking at the *concept* of being human.

Philisides has long been recognised as a figure for Sidney himself: Katherine Duncan-Jones calls Philisides 'Sidney's poetic persona', but the name was also used by Sidney in the Accession Day tilt of 1577, when he appeared as 'Philisides, the Shepherd Knight'.[59] In good humanist fashion Philisides is the name of both poet and soldier, maker and doer, and the fable he tells after the wedding of the shepherd Lalus and his love Kala has particular resonance within the current discussion. His beast fable is a warped creation narrative in which Man is created at the request of the animals to be their ruler. Man then exploits the generosity of the subjected animals and destroys the natural harmony. At the end the animals are told, somewhat dangerously, 'And you, poor beasts, in patience bide your hell, / Or know your strengths, and then you shall do well.'[60]

The fable has been interpreted in terms of its political meaning in two important critical readings: both David Norbrook and Annabel Patterson examine the fable and give different but complementary interpretations. Patterson traces the origin of the fable to Judges 9 and Aesop's *The Frogs Desiring a King*.[61] Norbrook emphasises the fable's acknowledgement of its fictional author, Hubert Languet, and sees it as 'a classic exposition of the radical Protestant fear of the growing power of absolute rulers.'[62] The fable is about the political system and the consensual nature of rule: literally, the animals made man, the subjects made their ruler.

The politics which can be traced in Philisides' fable and in *The Old Arcadia* as a whole can be found in a text which may have precipitated its completion. In a letter to the Queen in which he advised against the French marriage Sidney explicitly warns Elizabeth: 'These [subjects] therefore as their soules live by your happy government, so are they your chefe, if not sole strength.'[63] Sidney presents rule as neither mythic nor divinely ordained, but as at once consensual and exploitative: there can be power only if there are those without power. This dangerously reductive representation is at the heart of Philisides' fable. However, in *The Old Arcadia* Sidney does not only reiterate this idea, he also gives an explanation for writing it: he must not be silent. Unlike Gynecia who sees the dangerous truth of the monarch's desires but remains silent, Sidney sees the dangers of the French marriage and speaks out against it because he is, in his interpretation anyway, both a good reader and a true subject. In short, he is a human.

This political reading of Philisides' fable is available to the reader by looking in detail only at the underlying meaning of the piece. In doing this the reader is, with Patterson and Norbrook, an ideal, truly-human, humanist reader: the surface is dismissed as unimportant and the hidden (political) meaning is sought. But while it might seem plausible that the humanist reading should dismiss the animals in the fable, the animals can be read as offering up a vision of interpretation itself.[64] If we look at the surface meaning of Philisides' fable then we find a fable not about politics but about being human.

At the beginning of the fable the beasts exist without humanity in an original (prelapsarian) harmony, with, significantly in this humanist text, an original ability to speak. They decide they want a king. Jove unwillingly agrees and lends 'part of [his] heav'nly fire' to the creation, but orders that 'the rest yourselves must give'. Each animal donates what is clearly its emblematic quality:

> The fox gave craft; the dog gave flattery;
> Ass patience; the mole, a working thought;
> Eagle, high look; wolf, secret cruelty;
> Monkey, sweet breath; the cow, her fair eyes brought ...[65]

And each of the animals gives up its right to speak. They willingly relinquish the thing which humanism will later claim to be one of the defining features of the true human; animals are shown, once again, to possess the quality of human-ness before humans. In the fable, however, Man is created and goes on to abuse his power with his cruelty to the very creatures that made him. His cruelty is represented not only politically but by his ability to manipulate through language: 'Not in his sayings saying "I", but "we"; / As if he meant his lordship common be.'[66] Eloquence and rule have broken down irrevocably. Clarity of meaning and fairness of government come to an end together.

Sidney's fable not only echoes Aesop and Languet, it also inverts the ideas of Pico della Mirandola, a figure who seems to haunt Sidney's Reformed arcadian universe. In *On The Dignity of Man* Pico wrote: '[m]an fashions, fabricates, transforms himself into the shape of all flesh, *into the character of every creature.*'[67] This Mirandolan man, who celebrates his potentially beastly nature, is the antithesis of Sidney's fabled man who is made by the animals. Philisides' fable also reverses the pattern of creation which exists within the tradition where 'man is conventionally identified with this or that animal because certain animals are identified with particular human characteristics.'[68] Humanity is the *a priori* status in the writing of fables: it is, to return to Stallybrass and White, the always-already. In Sidney's fable, however, the human is presented as the creation of a number of emblematic qualities. The implication is that being human is not *a priori* but *a posteriori*, a creation of all that goes before. Sidney's man is an amalgam of all animals; more animal than the animals themselves. Patterson calls him 'the super-beast'.[69]

But, if we are using the terms of the didactic prefaces to contemporary poetry and fables, my reading, concentrating as it does on the surface narrative of Philisides' fable, is, of course, a bad one. It does not go beneath the surface to the 'real' meaning of the work. The implications of this bad reading are drawn out in Sidney's discussion of the reading of fables in *A Defence of Poetry* where he writes 'so think I none so simple would say that Aesop lied in the tales of his beasts; for who thinks that Aesop wrote it for actually true were well worthy to have his name chronicled among the beasts he writeth of.'[70] To read badly is to be a beast. Only the reader who fails to read correctly, who has forgotten their education, will find their closeness to the animal in Philisides' fable. Good, and by analogy human, readers will never have to face their animality because they have left it behind. They are not so much super-beasts as supra-beasts. Good readers are above the animal through their ability to look beneath the surface.

For Sidney a humanist education is inseparable from the Reformed ideas of Calvinism. There can be no transformation, no attempt at divine status (Basilius' first mistake), what there is instead is a fulfilment of potential. Where Martin Luther proposed that the faith of the infant is 'a divine possibility and creation in man, as opposed to a natural human possibility or work', so Sidney argues that the ability to interpret, to discern, is likewise a God-given possibility, even to postlapsarian man.[71] The always-already fallen nature of humanity is where the labour of humanism lies. Thus, where Pico cites endless possibilities, in Reformed thought such infinitude has gone with the Fall. In 1583 Stubbes wrote 'he that knoweth nothyng, is like a brute Beast. But he that knoweth al thynges (which thing none doeth but God alone) he is a God amongst men.'[72] This sums up, in many ways, the difference between the Mirandolan and the Reformed vision of the world. For Pico man can descend to the level of the beast (or, even worse, the bush), but can also ascend to the divine. Stubbes, however, allows for the former possibility – animal status – but denies the latter. Only God can be God, it was Adam's original sin to attempt to achieve a God-like and total knowledge which caused the Fall. In Sidney's romance the Oracle sits in the throne of Calvin's God and the characters enact a variety of responses to the world around them. There are possibilities – different interpretations – but there is also truth.

> Thy elder care shall from thy careful face
> By princely mean be stolen and yet not lost;
> Thy younger shall with nature's bliss embrace
> An uncouth love, which nature hateth most.
> Thou with thy wife adul'try shalt commit,
> And in thy throne a foreign state shall sit.
> All this on thee this fatal year shall hit.[73]

It is this divine truth which limits all of the actions of the text. The Oracle must be fulfilled, the King must commit 'adul'try' with his wife. But there can be judgement. The work of humanism – the act of becoming human – is hard, but it can be successful. But the pursuits of humanism which also serve to define it – education, the attainment of eloquence and wisdom – are pursuits which paradoxically both disclose the animality of humanity and emphasise the role of the animal in achieving human-ness. In the beast fable Sidney's Reformed humanism both reveals and ultimately relies upon the unnaturalness of humanity. Being human is learnt, it is not innate. But it is also, for Ben Jonson, purely theoretical.

IV

Sidney's injunction that poetry should both 'teach and delight' is central to Jonson's ideas about art.[74] In 'The Epistle' addressed to the two Universities of England which accompanies the printed text of *Volpone* Jonson writes, *'I have labour'd for* [men's] *instruction, and amendment ... which is the principall end of* poesie, *to informe men, in the best reason of living.'*[75] Like other fables, this one is being told for educative reasons. The 'Prologue' to the play speaks of 'ri'me, not emptie of reason', and of the poet's endeavour to 'mixe profit, with your pleasure'.[76]

It was not only the over-arching humanist endeavour which links Jonson to Sidney, however. Jonson follows Sidney's delineation of the interpretive community and reproduces his distinction of human and animal readers. In *Every Man In His Humour* Jonson wrote:

> Hang him, rooke he! why, he has no more iudgement then a malt-horse. By s. GEORGE, I wonder you'ld loose a thought vpon such an animal ... By his discourse, he should eate nothing but hay. He was borne for the manger, pannier, or pack-saddle![77]

Bobadill's attack on Squire Downright presents the human as animal and makes the distinction between the species in terms of wisdom and eloquence. The Squire's judgement can be traced through his discourse, and his discourse reveals his only birthrights to be those of the equine kind. The emphasis on the Squire's spoken word takes Sidney's idea of the human and the animal readers out of what can be termed in Jonson's thought the theoretical sphere of writing and into the practical sphere of speech. 'Speak that I may see you' is taken, by Jonson, to mean exactly that: spoken eloquence is the site of identity. Where Sidney's ideal reader, Euarchus, can assert that 'thoughts and conceits are the very apparel of the mind', for Jonson the mind is sought in the voice.[78]

It is for this reason that Jonson attempts to fulfil the humanist ideal of discernment in the theatre. Just as Sidney discovers himself to be the true human when he puts his theory into practice – writes his letter against the French marriage – so Jonson looks to the stage to find a place of eloquence and action, and it is in the interpretation of the drama that he traces Sidney's two categories of readers. Jonson divided theatre-goers into auditors and spectators. The auditor (from the Latin *audire*, to hear) listened to and pondered the meaning of the play, while the spectator (from the Latin *spectare*, to behold) merely watched, looked at the surfaces.[79] These two categories of theatre-goer reproduce for the stage the categories of good/human reader and bad/animal reader which are outlined in Sidney's work: Sidney's animal is Jonson's spectator. *'Language* most shewes a man' Jonson wrote: eloquence and identity were linked.[80] However, in transferring Sidney's theory into the practical realm of the theatre Jonson reveals the impossible nature of Socrates' dictum. In this world of spectacle there can be no interpretation. Just as Anthony Munday saw the frailty of the human senses of hearing and seeing, so Jonson, from a very different point of origin, also asserts human incapability in these terms. In Jonson's theatre there are only animals. There are no humans in drama.

The most obvious of Jonson's beast images appear in *Volpone*. The play has been read as a dramatisation of the traditional fable of the cunning fox.[81] This is, of course, true. But the text also utilises animal imagery in a way which is far more complex than this. We are not dealing only with Adams' *'as Lambs'* here. The play begins with what might be termed two dramatic introductions: first, Volpone's eloquent worship of his gold; and second, Mosca's interlude. In the former God is displaced by wealth, and in the latter the journey of Androgyno's soul represents an emblematic recognition of the closeness of the human and the animal which dismisses Sidney's emphasis on Jove's donation of the 'heav'nly fire' (the soul) to the making of man.[82] We are, in Volpone's bedroom, in a world without God and without species difference. The main action of the play truly gets underway with the stage direction *'One knocks without'*. 'Give me my furres' says Volpone, and the fable begins.[83] The actor dresses up for the part.[84]

Volpone controls the action of the early part of the play before he is superseded by his servant. He is, in many ways, the author of his own performance. He performs and gives directions, or so he thinks. The real work, the real crying of stage tears, however, is done by the actor Mosca. This is an image of writing: Jonson presents a playwright who loses control. Volpone's sense of himself as in command is Jonson's mockery of his own role as controlling author. The humanist becomes a playwright, and the theatrical parallel is found in the image of the man putting on furs: an indebtedness to animals emerges in both. The theatre-goers, Jonson's clients, will be taken in but ultimately will misinterpret, just as Volpone's clients are taken in and misinterpret the vision of death they see before them:

> Now, now, my clients
> Beginne their visitation! vulture, kite,
> Rauen, and gor-crow, all my birds of prey,
> That thinke me turning carcasse, now they come.[85]

But at the end of the play the Avocatori, the representatives of (a rather dubious) justice, prevail and the beasts are punished: 'Let all, that see these vices thus rewarded, / Take heart, and love to study 'hem'.[86] The message is clear, but in its overwhelming clarity it is acknowledged as a humanist failure: like *The Shepheardes Calender* and unlike *The Old Arcadia*, there is no work to be done by the reader here. Where humanist educationists asserted the importance of repetition, of labour, Jonson does all of the work for the theatre-goer.[87]

The failure of the humanist project in the theatre is reiterated in 'The Epilogue' of *Volpone* where entertainment is shown to overwhelm the play's morality. The newly condemned title character steps towards the theatre-goers:

> Now, though the FOX be punish'd by the lawes,
> He, yet, doth hope there is no suffring due,
> For any fact, which he hath done 'gainst you ...[88]

The laws of the play and the morals of the drama diverge and the Fox gains his applause; the viewers have forgotten the simple lesson. All of Jonson's fears about the theatre seem to be justified: the entertainment of the spectacle has overrun the profitableness of the text. The animal interpreters in the theatre have had the moral spelt out for them and it is easily forgotten; have had the love of the theatrical fulfilled, and visual pleasure has ousted wisdom. There is no bridling or breaking-in: there is only entertainment. There are, to return to Socrates, no listeners only watchers: humanism is inverted.

In this interpretation the dressing up of Volpone seems like a parody of the actions of Machiavelli, one of the most infamous humanists in early modern English culture.[89] In a letter to Francesco Vettori dated 10 December 1513 Machiavelli wrote:

On the coming of evening, I return to my house and enter my study; and at the door I take off the day's clothing, covered with mud and dust, and I put on garments regal and courtly; and re-clothed appropriately, I enter the ancient courts of ancient men, where, received by them with affection, I feed on that food which only is mine and which I was born for ...[90]

No direct link can be made between Jonson and Letter 137, but Machiavelli's moment of humanist epiphany could easily be a direct reversal of everything that happens in *Volpone*.[91] Volpone only leaves his room when he is forced to; he puts on not the clothes of a courtier but the costume of an animal; he is only received with falsity, there is rape and no affection in the play; and finally, Volpone does not feed, but rather is fed upon by the parasite Mosca. Volpone is the anti-humanist.[92] To add to the desire to read the play alongside Machiavelli's most well-known letter, in it Machiavelli – who was frequently termed the Fox – writes with self-mockery about his life in the country: 'I have until now been snaring thrushes with my own hands'.[93] The image of the fox pre-tending death in order to catch the birds of prey who swoop down on his supposed carcass is replicated in *Volpone*, and can be traced in medieval bestiaries. In one bestiary this story is appended with 'The Devil has the nature of this same.'[94] Machiavelli mocks his detractors and presents the activity of the Antichrist, catching thrushes, as his own country pursuit.

But unlike Machiavelli, the self-mocking humanist, Volpone is not utilising the fable to learn more about the operations of the world; instead, he *is* a fable. The mighty eloquence which Volpone does display links him to Sir Epicure Mammon, a fictional character in *The Alchemist*, rather than to Ben Jonson, the humanist. On close inspection the fur-clad fox is not a humanist at all, and is certainly not the author. What Jonson is doing throughout *Volpone* is making the role of the interpreter impossible. The audience is drawn to the character who displays eloquence and who controls the play. Volpone is the self-conscious imitator of the beast: his is not a literal metamorphosis, but, to use Thomas Adams' phrase, it is playing *as* a fox. He puts on his costume, the other characters are always-already in theirs. Recognising this the audience seem to have exercised their human-ness, seem to have looked beneath the surface, the furs, to see the always-already human. But, of course, if the audience know their Socrates (the link between listening and literacy is a deliberate one) they will know that the eloquence of the central character is where his true character is to be sought, and Volpone's eloquence when interpreted is perverse: his first act is to displace the divine with wealth: God with Mammon.

This interpretation seems to leave the auditor in a position of power, of human-ness; they have seen through the surface to the deep meaning. 'The Epilogue' confuses things, however, as it forces the audience to applaud the pervert. They are seduced by false eloquence and are revealed to be spectators after all. Just as Bobadill seems to exercise his humanity in his assertion that Squire Downright judges like a malt-horse but is ultimately revealed to be as animal as the Squire, so the audience seem to be judging the spectacle of *Volpone* but are in reality merely spectating. There can be interpretation, but there is still animality. Douglas Duncan argues that in *Every Man In His Humour* '[m]ost of the characters criticize each other, but lack the moral equilibrium of achieved self-hood, and so have no right to pass judgements at all, their critical faculty having failed in its primary task.'[95] The critical faculty of the theatre-goer also fails: in *The Alchemist* Jonson writes mockingly of 'Iudging Spectators', an impossible status.[96] It is the reader who is Jonson's ideal and only interpreter.

In publishing his *Works* in 1616 Jonson placed himself alongside Aristotle, Plato and Socrates, and underlined a link between the classical past and the present. Both ancient and modern writers needed to be interpreted deeply in order that the reader could understand. Unlike the 'Owles and Popiniayes, and Rauens, and pyes' of the homily, and the 'vulture, kite, / Rauen, and gor-crow' of *Volpone*, the reader has discernment.[97] Where the spectators applaud the fox at the end of *Volpone*, the reader, lacking the visual exhibition of the bow, can see the danger of this moment. But the reader does not encounter speech, and the notion of speaking and seeing breaks down: reading and understanding is a different process. Jonson's theatre, the space in which he practises Sidney's theory of human-ness, reveals that theory to be purely theoretical and not real. Just as Holinshed, Dekker and Lupton propose that being human is impossible in the Bear Garden, Jonson proposes that, in the theatre, the space which truly enacts the humanist emphasis on eloquence and action, watching confirms the status of humans as animals.

In the second century AD Aulus Gellius defined *humanitas* as 'education or training in the liberal arts'. Such training led to increased wisdom and eloquence through imitation of the classics. But Gellius went on to say that '[t]hose who earnestly desire and seek after these are most highly humanized [*maxime humanissimi*].'[98] There is human, there is earnestly desiring human and then there is *humanissimi*. There is no such thing as human as a single defineable substance. The human is actually being made truly human by humanism. There are always multiple versions: some humans are more human than others. The animal, like the first instance of the human in Gellius' account, and like the birds who sing without understanding, has no desire. English humanists like Sidney and Jonson propose the possibility of 'achieved self-hood', something which comes with desire, but they also see the stronger possibility of the absolute opposite. In humanism, the animal is still a potential status for humanity.

V

For Jonson the dangerously visual nature of the theatre and theatricality was frequently represented in the human body. Volpone's on-stage dressing-up and putting on of make-up, Mosca's rich attire, Voltore's advocate's garb, all enact the physicality of disguise and the use of the body to prevent true judgement. Even Sir Politic Would-Be thinks that disguising his physical form will prevent his discovery.[99] Seeing through these costumes, as Euarchus was able to do in *The Old Arcadia*, enacts discernment. But this discernment was lost in the theatre. The implications of theatrical disguises are not found only in Jonson's humanism, however, the body, as noted, was also significant in Reformed theology.

This is where a link can be traced between Sidney, Jonson and anti-theatrical writers. The question of visibility (surface) which is at the heart of Sidney's notion of judgement and Jonson's fears about the stage is also central to the Reformed attacks on the theatre of the late sixteenth century. The issues of watching, interpreting and the body which these ideas throw up are brought together in one moment of horror in William Prynne's 1633 text *Histrio-Mastix*. It is in the animal once again that the dangerous possibility of the spectacle – the surface – is played out.

Examining the construction of the subject in anti-theatrical works, Laura Levine notes the fragility of status and writes '[n]o one seems to have any inherent identity and everyone can be converted into someone else'.[100] The conversion she sees is into a new being, is about a loss of personal identity. For William Prynne, however,

Levine's some*one* else needs to be translated. Prynne's work seems to undermine the sense of the separation of the species which is inherent in the notion of the always-alreadiness of human status. When Prynne proposes transformation he is proposing some*thing* else: the achieved-self is animal.

In *Histrio-Mastix* Prynne, like so many anti-theatrical writers before him, launches an attack on '*mens disguising of themselves like women*' in the theatre. Following the Deuteronomic code (Deuteronomy 22: 5), cross-dressing is regarded as unlawful among Christians: 'there is no warrant at all in Scripture for any such Stage-disguises, but very good ground against them.'[101] However, the attack on transvestism is not the end of Prynne's diatribe: he writes 'how much more then mens transfiguring of themselves into the shapes of Idols, Devils, Monsters, Beasts, &c. betweene which and man there is no Analogie or proportion, as is betweene men and women.' He goes on, arguing that scripture – that central agent of Reformed ideas – 'commands men, *not to bee like to Horse or Mule, which have no understanding*: therefore not to act their parts, or to put on their skins or likenesse.'[102] For Prynne dressing and being are absolutely linked: costume – fur – as Perkins, Stubbes and Jonson proposed, epitomises the mind of the wearer.

Prynne's reading of the Bible at this point in the text moves from the figurative to the literal. In scripture men are commanded not to act like animals, that is, not to behave in a way which resembles animal behaviour (see, for example, Psalm 32: 9 which Prynne quotes directly in his polemic), nor to dress in animal skins like the deceptive Jacob (Genesis 27: 16). The injunction against acting the parts of beasts in the theatre is, significantly, Prynne's own, there is no scriptural precedent. Prynne continues:

> And must it not then bee mans sinne and shame to act a Beast, or beare his image, *with which he hath no proportion*? What is this but to obliterate that most *glorious Image which God himselfe hath stamped on us*, to strip our selves of all our excellency, and to prove worse than bruits?[103]

It is the destruction of the species barrier which poses the greatest threat to the status of humanity. But this threat to the separation of the species is figured as the *rational* extension of biblical laws:

> that God, who *prohibits, the making or likenesse of any beast, or fish, or fowle, or creeping thing, whether male or female* ... must certainly

condemne the putting on of such brutish Vizards, the changing of the glory, the shape of reasonable men, into the likenesse of unreasonable beasts and creatures, to act a bestiall part in a lasciv-ious Enterlude.[104]

Prynne does not need direct scriptural authority to make his point, the italicisation (symbolising quotation from the Bible) has stopped. God 'must' abhor cross-species-dressing, he argues; there is no proof, just reason. Prynne goes against Calvin's warning about examining the 'hidden recesses of the divine wisdom' and declares that he knows what God thinks, and goes against other anti-theatricalists who prided themselves on the scriptural basis of their arguments.[105] Anthony Munday proposed

that I maie seeme to write nothing without ground, or to finde fault without cause, I will, GOD to freend, set downe nothing to prooue mine assertion good, but what scripture shal warrant, ex-amples confirme, reason allowe, and present experience ratifie.[106]

Munday argues conventionally that scripture is primary and all human reason merely secondary. Prynne's reason, however, takes more than a leap of faith, it takes a leap of logic. Where Jonson, applying logic presents Volpone, the wearer of furs, as the anti-humanist, so Prynne presents the man who dresses as an animal as the anti-human. Stubbes regarded leather as the original and pure form of clothing given by God to Adam and Eve which 'should be as a rule or pedagogie vnto vs, to teache vs that we ought to rather walke meanly, and simplie, then gorgeouslie, or po[m]pouslie'.[107] Prynne, on the other hand, sees the animal-costume as being more like Stubbes' depiction of fashion, as evidence not merely of the Fall, but of the animality of humanity.[108] The dangers of the visible return in the shape of the animal. Reformed ideas of regeneration and the humanist distrust of the surface are enacted in the animal costume. Clothing is not evidence of an inner state, it is transforma-tive: to wear is to be.

Prynne's assault on the difference between the human and the animal brings together the emphasis on the body in Reformed ideas and the mind in humanism.[109] If sight is the predominant sense and the visual elements of society such as clothing and the theatre can deform the body then there seems to be a real limitation on the places in which the human can be found. Reading is the theoretical space where human-ness can be constucted, but reading is limited to the literate, and is a solitary act which removes the reader from the speech community which humanism seems to emphasise.

Joanna Martindale has suggested that humanism should be defined 'by its characteristic pursuits and attitudes', and this is a useful definition here.[110] Humanism is a process and not an object: it is about expression, interpretation and not being. Achieving human status, becoming *humanissimi* is the thing itself. The implication is that being human is, paradoxically, outside humanism and it is through the animal that human-ness can be found. This lays bear the problem. There is no human without an animal present, but the presence of the animal can itself disrupt the status of the human. The failure of humanist logic to assert a difference between human and animal does not mean that we must abandon reading altogether. In another form of reading, the deep reading of the animal in Francis Bacon's writing, another notion of the human is asserted.

4

Seeing All Their Insides: Science, Animal Experimentation and Aesop

In *The Advancement of Learning* Francis Bacon wrote 'it is not good to stay too long in the theatre.'[1] Where for Reformed thinkers the spectacle of the theatre displaced judgement and watching was a dangerous thing, and for Jonson the humanist looking was not a threat to the human but was the thing which actually proved the human an animal, Bacon also viewed the theatre with suspicion. But for him the theatre was a threat for a very different reason: because of its fictitiousness. The visual nature of the theatre which so disturbed the human in Reformed and humanist ideas was, in fact, vital to the development of the new science and to the expansion of the power of humanity which the new science proclaimed. The truths of science were only enacted through the sight: Bacon wrote 'For I admit nothing but on the faith of eyes, or at least of careful and severe examination; so that nothing is exaggerated for wonder's sake, but what I state is sound and without mixture of fables or vanity.'[2] The visual is at the heart of the new science: to see is to believe.

Fiction was, however, another matter. At the heart of what might be called this third form of anti-theatricality (Baconian as opposed to Reformed or humanist) was the question of representation. The playhouse stage represented reality in 'unreal and scenic fashion'.[3] The endeavour of the new science was to eradicate the layers of fictitiousness which existed in what Bacon called 'fables and superstitions and follies' and reclaim the real.[4] Fables lay between the human and a true understanding of the world around them and their removal was vital because Bacon was looking for a new way of engaging with the materiality of existence.

This chapter moves then from the Jonsonian theatre to the anatomy theatre and begins with the representation of animals within what has been termed in hindsight the old science. A close examination of the content of works from the Middle Ages and the early modern period reveals a sense of both continuity and change. What remains in place in the early modern texts is the notion of the absolute centrality of humanity, what has changed are the means of enacting that centrality. Bacon asserts that he is taking on the changes and getting rid of the hang-overs of the old science but the re-appearance of myths in his work reveals a paradox in the theory which is extended through an analysis of the practice of the new science.

Bacon's empiricism proposes a return to the prelapsarian state which is interpreted as being based on a dismemberment of both paradise and, of course, animals. Looking beneath the surface of the animal, the deep reading of the beast which takes place in experimentation, is presented as the fulfilment of the ideal which failed in humanism. However, the reading of the animal in vivisection repeats the paradox of humanism's reliance on the animal. The absolute animality of the human body is revealed even as humanness is expressed. As such the destruction of the boundary which exists between human and animal continues. Reformed ideas, humanism and the new science are all linked by their difficulties in sustaining an idea which separates the species.

But Bacon proposes a solution to this problem. The vivisector, the Baconian scientist, is presented as the good reader, the true human. Cutting up animals becomes the thing which reveals both the animality of the human body, and the human-ness of the human. This contradictory status of the human is then traced through to Bacon's final proposal for his scientific endeavour, the fabulous travel narrative, *The New Atlantis*. When he published this unfinished tale in 1627 William Rawley prefaced the text with a note 'To The Reader': 'This fable my Lord devised, to the end that he might exhibit therein a model or description of a college instituted for the interpreting of nature and the producing of great and marvellous works for the benefit of men'.[5] Just as the human exercises human-ness through experimenting upon animals and reveals in this experimentation his image in the beast, so Bacon can only dismiss fables in a fable. The scientific truths which empiricism claims for itself can only be expressed through the things which it discounts. But this can be taken even further. The human of Bacon's philosophy finds a precursor in the life of the most famous fabler of all: Aesop. The fables which give the animals voices, which were so central to and problematic within humanism, are not left behind by the new science, they epitomise it. In fact, Bacon's human emerges as an echo of the Aesop-animal. The difference between human and animal is, even in the new science, a myth which is based on faith rather than proof. The chapter begins, as it ends, with the things which empiricism says we need to leave behind; the fable.

I

URSUS the Bear, connected with the word '*Orsus*' (a beginning), is said to get her name because she sculptures her brood with her mouth (*ore*) ....
... The males respect the pregnant females with the decency of a private room, and, though in the same lairs for their lying-in, these are divided by earth-works into separate beds.[6]

The twelfth-century bestiary presents the bear first through the origin of its name; and second through its likeness to the human. The animal is meaningful in two ways: it offers a fabulous image of creation ('licking into shape'), and it also reveals something about human mores. The bestiary narrative makes language concrete – Ursus is related to *Orsus* and *ore* – and works to naturalise cultural activities; the lying-in in which the pregnant woman is secluded is found to be a *natural* part of childbirth. The anthropocentrism of the bestiary is clear: animals are studied because they allow us to say things about humans and human lives.

Similarly, the twelfth-century naturalist Alexander Neckam argued that 'all created things ... were made to be obedient to man or to serve him' and in the thirteenth century Albertus Magnus presented the natural world as a source of human health: 'haemorrhoids goeth away from him, which sitteth upon the skin of a lion.' 'The tooth of a mare put upon the head of a man, being mad, delivereth him anon from his fury.'[7] The centrality of humanity evidenced in these texts is repeated in the next century in the work of Bartholomeus Anglicus (c.1360). In Liber XVIII 'De Animalibvs in Generali' Bartholomeus returns to the fourth-century work of Basilius to emphasise his main point:

[Basilius] calleth tame beastes *Iumenta*, and sayth, that they be beastes graunted and ordeyned to vse and to helpe of mankinde.

And some be ordeyned to trauaile, as horses, Oxen, and Camells,
and other such: and some to beare wooll for clothing of men, as
sheepe and other such, and some to be eaten, as swine and
pigges ...[8]

The animals are divided into categories which represent the functions they fulfil in human society. The sheep's purpose is to carry
the wool until it is needed by men.[9] It is the work of knowledge to
underline this.

To call these texts scientific is obviously very problematic. Based
upon myths, fables and absolute and unquestioned anthropocentrism, they seem very different to what we understand science to be
now.[10] But these medieval texts offer a basis for this new understanding of science. An examination of the process of change
reveals the ways in which links can be made between the bestiary
and Bacon. As a starting point a reassessment of the methods of the
bestiaries reveals a similarity to the new scientific emphasis on
seeing. Historian of science Charles Raven is particularly dismissive
of the bestiary. He writes of them that:

> the interpretation of nature consisted not in reading out of it the
> lesson of its true character but in reading into it an elaborate
> arbitrary and artificial significance. It ceased to be a means for
> discovering the character of reality, and became irrelevant and
> negligible except in so far as it was the source of imagery and
> anecdotes for the moralist and preacher.[11]

The terms 'arbitrary', 'artificial' and 'moral' are echoes of the language which Bacon will use to describe the old science. Raven's
evaluation is overlaid with the language of the new science in a way
which fails to recognise some of the more detailed changes which
were taking place in this archaic form.[12] There were 'halting steps'
towards realism in the representation of animals in the bestiary, and
the more realistic representations were of those animals which
would have been observed on a daily basis in England.[13] Even in
the Middle Ages a sense of real observation, as opposed to the mere
repetition of myths, was emerging which offers a link to the
scientific ideas of the early modern period.

Another, and perhaps more signicant, link can be traced when
the more modern works are read alongside the bestiaries. What
becomes clear is that the sense of absolute difference between the

old and the new is unworkable.[14] A number of the medieval texts were reproduced throughout the early modern period: Stephan Batman, for instance, translated the work of Bartholomeus Anglicus in 1582, and nine editions of Albertus Magnus' *Book of Secrets* appeared between 1550 and 1637.[15] Claims for a shift in knowledge in the sixteenth century ignore the continuing popularity of these texts. Indeed, the editors of *The Book of Secrets* have gone so far as to write of the 'fundamentally medieval cast of mind of the Elizabethan reader'.[16]

Similarly, when Edward Topsell's *The Historie of Foure-Footed Beastes* appeared in 1607, two years after the appearance of Bacon's *Advancement of Learning*, there remained within it mythical hang-overs from the Dark Ages: the unicorn is listed, among other animals.[17] The echoes of earlier texts in Topsell's work can also be traced in another way. In 'The Epistle Dedicatory' he wrote that Conrad Gesner's *Historia Animalium* (1551–8), his main source, intended 'to shew vnto men what Beasts are their friendes, and what their Enemies, which to trust, and which avoyd, in which to find nourishment, and which to shun as poison'.[18] Animals are only important in the functions they fulfil in a very human world.

Topsell's collection of descriptions, fables, and tales of animals begins with 'The First Epistle of Doct. Conradvs Gesnervs before his History of Foure-footed Beastes, concerning the vtility of this story'. In this epistle Gesner sets out his reasons for his investigation of the natural world:

> the knowledge of this naturall part of Phylosophye, is very neces-sary and profitable to Physique, and that many waies; First, for that many beastes are vsed for meate, nourishment, and medi-cine, and for that cause are not only applied outwardly, but inwardly to the body of man ...[19]

Animals need to be understood so that we can consume them with more ease. They are for human use. This is a return to the sheep as the wool-carrier for humanity. Gesner also returns to the fable, and writes that men can take 'instructions out of beasts, by imitation of whose examples, the liues and manners of men are to be framed to another and better practice'.[20] His endeavour exists alongside rather than as a replacement of the medieval works.

However, while Gesner's rationale may replicate the medieval model his practice reveals important changes. It brings the humanist emphasis on the return to the sources together with the new scientific belief in actual observation and reliable eye witnesses.[21] He writes

> I did not only sit still and turne ouer books, but gaue my selfe
> diligently to enquire of euery Country-man or travailer, a perticu-
> lar and exact obseruation of the nature of euery beast: and for this
> course I conferred with strangers of other nations, which by any
> occasion either were resident among vs, or passed accidently
> through our country, & made of their relation the most diligent
> notes that I could gather, conferring them with olde writers, and
> comparing one with another, till I had digested the perfection of
> my intented story, and the method thereof for the drawing of it
> into a perfect body.[22]

In his method we can trace both the 'olde writers' who were so important in the establishment of facts in the medieval works, and a sense of experience and enquiry which is was emerging in the increasing naturalism of the bestiaries. Gesner's work represents a mid-point in the development of science.

Another alteration, this time in the changes in the text of Albertus Magnus' *Book of Secrets*, also underlines the slow process of change which was taking place. The prefatory letter to the 1550 edition of Albertus Magnus' text reads:

> Sith it is manifestly known, that this book of Albertus Magnus is
> in Italian, Spanish, French, and Dutch tongues, it was thought if it
> were translated into the English tongue, it would be received
> with like goodwill and friendship, as it is in those parts.
> Wherefore use then this book to mitigate and alacrate [enliven]
> thy heavy and troublesome mind, as thou hast been wont to do
> with the book commonly called the Book of Fortune: for, believe
> me, whatsoever is promised in either of them both, this or that, is
> alonely to this end.[23]

This letter is repeated in the 1617 edition of the text but a phrase is added at the end: 'I referre thee to the tryall of some of his secrets, which as thou shalt find true in part or all, I leave to thine owne report or commendation.'[24] Even as this medieval work is being reproduced there is a new understanding of science, an expectation that to be accepted as 'true' in 1617 experimentation must take place.

This pattern of the development of science – the continuation of the perception of the purpose of the animal, and the alteration of the method of understanding that purpose – is underlined in a later passage in Gesner's 'Epistle' where an idea which is at the heart of the new science emerges: 'it is more safe and without impiety, to make tryall of a new medicine upon a beast, rather than vpon a man.'[25] Animal experimentation was not, of course, invented in the early modern period, but as Richard D. French has noted, following the publication of Harvey's *De motu cordis* in 1628, 'for the first time the practice of experimentation upon living animals impinged on the consciousness of the educated elite in general.'[26] The importance of animal experimentation, which is so central to the new empirical method (I will return to this), is born out of an absolute and unquestioning sense of anthrocentrism which comes from the old science but is developed in a very different way in the new. In her important feminist revision of Bacon and seventeenth-century science, Carolyn Merchant characterises the movement as being from regarding nature as a teacher, to seeing nature as a slave.[27] But, just as a historian of science such as Charles Raven reproduces without question Bacon's language in his dismissal of bestiaries, there is also a refusal to look for inconsistencies in empiricism. Even Merchant does not really allow for any but rather examines Bacon's 'new ethic sanctioning the exploitation of nature.'[28] This ethic is certainly central to the new science, what is also significant, however, is the way in which Bacon arrives at it. By taking the changes which can be traced in the increasing emphasis on experience, proposing the abandonment of the fables of the old method and yet maintaining the centrality of humanity Bacon establishes, it would seem, a truly new and empirical science. The reappearance of myths in Bacon's *Of The Wisdom of the Ancients* makes such a claim, however, problematic and the paradox of the myth at the heart of the new science is where I begin to question the implications of Bacon's philosophy and to examine the differentiation of man from beast and the establishment of an unproblematic category which can be termed human in the developing empiricism of early modern England.

II

Bacon's relationship with the fable seems to remove his notion of science from the fictions found in medieval work. In 'The Plan' of

The Great Instauration he remarks that 'fables and superstitions and follies which nurses instil into children do serious injury to their minds'.[29] The vulgar (it is the nurse and not the parent who passes on the fables) imprint vulgar ideas on the formative mind and the damage is almost irrecoverable:

> No one has yet been found so firm of mind and purpose as res-
> olutely to compel himself to sweep away all theories and
> common notions, and to apply the understanding, thus made fair
> and even, to a fresh examination of particulars. Thus it happens
> that human knowledge, as we have it, is a mere medley and ill-
> digested mass, made up of much credulity and much accident,
> and also of the childish notions which we at first imbibed.[30]

A similar idea had been stated earlier in *Thoughts and Conclusions on the Interpretation of Nature or a Science of Productive Works* where Bacon wrote that 'Infants as they learn to speak necessarily drink in a wretched hotch-potch of traditional error. And however much men … advance in wisdom and learning … they can never shake off the yoke.'[31] Just as both Reformed thinkers and humanists pro-posed that the child must be taught from the earliest possible age – when it only just has the capacity to understand in some cases[32] – so Bacon emphasises the significance of that early education, but for very different reasons. The gap between humanist notions of edu-cation and Bacon's ideas is an important one.[33] He advocates a deliberate and conscious move away from teaching the very things which are at the heart of the humanist education; fictions and fables. The reasons which he gives for his attack on fables are simple: the 'Idols of the Theatre', as he calls received ideas in *Novum Organum*, represent reality like the playhouse stage.[34] This fictitious representation lacks power and it is the job of the scientist to under-stand and control nature as it exists in reality not in fables. The scientific endeavour must be based on direct observation for it to have any real meaning. If other methods are used natural history becomes 'as unstable as water and as gusty as wind'.[35] The strength, stability and, importantly, materiality of science vanishes if fables are a part of the project.[36]

But it is not merely the rejection of the moralising qualities of the old science which characterise Bacon's endeavour. The rejection of the sense of the inherent truth of ancient learning is also important. Where Albertus Magnus and Bartholomeus Anglicus rely upon the classical writers – 'Aristotle saith' – to make their proofs, Bacon writes 'generally speaking science is to be sought from the light of nature, not from the darkness of antiquity.'[37] The so-called ancients are reinterpreted as the exact opposite, and the moderns are represented as the real ancients: 'For the old age of the world is to be accounted the true antiquity; and this is the attribute of our own times, not of that earlier age of the world in which the ancients lived.' This, however, is merely academic because 'truth is to be sought for not in the felicity of any age, which is an unstable thing, but in the light of nature and experience, which is eternal.'[38] It is the method and not the historical period which gives a work its real significance. Experimentation will prevent the exaggeration and myth-making which has occurred in earlier scientific work because it offers up nature as it exists. This is where Gesner's use of travellers' recollections pre-empts, and the addition of the clause about testing which appears in the 1617 edition of Albertus Magnus' *Book of Secrets* reflects, the changing notion of science. There is a humanist return to the source, but there is also, in opposition to humanist ideas about education, a sense that trusting to memory is no use: direct experience is necessary.

Bacon would therefore appear to be proposing a straightforward dismissal of fictions and fables in his theory. However, in *Of The Wisdom of the Ancients* he presents a very different view. In this work he argues that:

> the truth is that in some of these fables, as well in the very frame and texture of the story as in the propriety of the names by which the persons that figure in it are distinguished, I find a conformity and connection with the thing signified, so close and so evident, that one cannot help believing such a signification to have been designed and mediated from the first, and purposely shadowed out.[39]

Fables bear a hidden truth which Bacon is able to uncover. This interpretation is an overturning of the account of the role of fables which appeared in *The Advancement of Learning* where he argued that:

> I do rather think that the fable was first, and the exposition devised, than that the moral was first, and thereupon the fable framed ... but yet that all the fables and fictions of the poets were but pleasure and not figure.[40]

For Paolo Rossi this change of heart is to be regarded as an evolution of Bacon's ideas.[41] Such an interpretation, however, fails to
acknowledge Bacon's reassertion of the unscientific nature of
mythic ideas which appear in works which follow *Of The Wisdom
of the Ancients*; not least in *Preparative Towards a Natural and
Experimental History* where Bacon dismisses them as 'slight and
almost superfluous'.[42] To search for an unequivocal development in
Bacon's thought is more problematic than Rossi allows.

In a more recent study Timothy H. Paterson offers another interpretation which posits that *Of the Wisdom of the Ancients* is 'radically
insincere' and that Bacon's 'belief in the real existence of such
"ancient wisdom" [is] wholly and deliberately feigned.'[43] This, as
Paterson notes, is something which most critics might shy away
from: 'the very possibility of radical authorial irony or insincerity' is
a direct attack on our notion of the author as a source of meaning.[44]
But Paterson's notion of insincerity actually places Bacon alongside
Sidney and Jonson. Ie too makes interpreting the meaning
difficult: an application of reason is required.

In *Of the Wisdom of the Ancients* Bacon interprets classical myths as
revealing meanings which support his scientific endeavour, but he
also argues that the fable 'will follow any way you please to draw
it', that 'meanings which it was never meant to bear may be plaus-
ibly put upon it'.[45] Such contradictions lead Paterson to claim that:

> it seems to me simpler, far more plausible, and more consistent
> with Bacon's obvious stature as a thinker to assume that he
> always meant what he said in speaking of the pretended exist-
> ence of ancient wisdom as primarily a means of adding prestige
> to his own thoughts through a conscious deception, and that he
> wrote *Wisdom of the Ancients* intending precisely that deception of
> many of his readers.[46]

Recognising the continuing power of ancient ideas, Bacon interprets them to back up his own ideas about the role of natural
history in society. In this text fables literally 'serve to disguise and
veil the meaning, and they serve also to clear and throw light upon
it'.[47] 'Infantile' notions are presented as both useless and useful, and
it is for the reader to recognise and interpret the paradox. *Of The
Wisdom of the Ancients* becomes a humanist, rather than a Baconian
text. In emphasising the act of reading fables Bacon takes the reader
away from the thing which should be read – nature.

If Paterson's notion of the irony of *Of the Wisdom of the Ancients* is correct then this text must be read as a contradictory document: it contains an acknowledgement of the pervasiveness of the learning of the ancients and of fables; a recognition of interpretation as a humanising skill; and, paradoxically, a call to move away from such poor science. Bacon recognises humanism but discounts it with an ironic flourish. Paterson cannot but offer this reading of the work because he wishes to view Bacon's works as consistent: he is, after all, a thinker of 'obvious stature'. I want to argue that the problems in asserting the coherence of Bacon's relation to the myth reflect problems in his construction of species difference.[48] Just as the myth resurfaces in the new science even as Bacon declares its dismissal, so the animal returns to undermine the status of the human.

III

The practical, as opposed to theoretical, aim of the new science was a straightforward one: 'to stretch the deplorably narrow limits of man's dominion over the universe to their promised bounds'.[49] Bacon was attempting to return humanity to its original prelapsarian position. In Genesis 1: 28 God gave Adam 'dominion over the fish of the sea, and over the fowl of the air, and over every living thing that moveth upon the earth.' This religious basis of the new science is important: we are forced to recall that Bacon's natural philosophy was not a wholly this-worldly endeavour.[50] The link between Bacon and the Bible is also a reminder that he saw limits to the operations of the new science. The search for knowledge was what caused the Fall in the first place and Bacon's endeavour was to return 'man's dominion ... to [its] *promised* bounds'. Science is not discovery, but a recovery of what has been lost.

In *Valerius Terminus of the Interpretation of Nature* Bacon wrote that 'being in his creation invested with sovereignty of all inferior creatures, [man] was not needy of power and dominion'.[51] The original order of the universe – while described in Genesis as 'dominion' – was, for Bacon, sovereignty.[52] Meat-eating was a postlapsarian phenomenon, as was the wearing of animal skins.[53] The model of power is very different after the Fall: Adam earns his bread through work, 'in the sweat of [his] face' (Genesis 3: 19). Part of the work which is introduced after the Fall is the control of the animals who have lost their original innocence and become wild. Henry Holland

wrote that man's 'soveraigne dominion, and rule, and power ...
ouer all creatures' was one of the clearest representations of human
perfection in paradise. After the Fall, however, this perfection vanished: 'for the creatures all repine, disdaine, and grone to serve
him.'[54]

Adam and Eve's desire for knowledge caused the animals to
become wild and antagonistic towards humanity. Bacon's new
science was a form of learning which would retrieve the human
control over the natural world. It proposed a return to the moment
of Edenic perfection when human power over the animals was
absolute. The new science, unlike the old, was not a fictitious return
of human power through myths and fables which, with their multiple meanings, were like the searching for truth under the shadow
of the Tower of Babel. It was an actual return to real power. Bacon
dismissed other forms of learning and notions of progress, and
proposed:

> therefore it is not the pleasure of curiosity, nor the quiet of resolu-
> tion, nor the raising of the spirit, nor victory of wit, nor faculty of
> speech, nor lucre of profession, nor ambition of honour or fame,
> nor inablement for business, that are the true ends of knowledge;
> some of these being more worthy than other, though all inferior
> and degenerate: but it is a restitution and reinvesting (in great
> part) of man to the sovereignty and power (for whensoever he
> shall be able to call the creatures by their true names he shall
> again command them) which he had in his first state of creation.[55]

Bacon reveals the failings of all other forms of human dominion –
religious, humanist, and economic – and cites the new science as
all.

Naming is central. To name in Bacon's philosophy is to exercise
power. The prelapsarian authority of human over animal was a
figure of God's authority over humanity. The Fall led to a distancing
of God from man just as it led to a distancing of man from animal.
Writing in 1607 Henry Arthington proposed that the resemblance
between God and man was vital and fragile;

> So that in these was no dissent,
>   Twixt God and Man, (for gifts most cleare)
> Saue (*all in God, were permanent*)
>   But man might change, (as did appear).[56]

Thomas Morton represents Adam's prelapsarian status through the idea of speech: 'man before his fall liued thus happily in the presence of God, with whome he had daily, and as wee may say, familiar conversations'.[57] As these dialogues cease, so does the communication with the animals. Humans no longer understand animals – call them by their true names – and conversation breaks down (communicate and community share a common root).

Adam's naming of the beasts is an exercise of human power over the animals of huge theological importance. Calvin sees this moment as a recognition of absolute human understanding: 'he named them not at a venture, but of knowledge he gaue to euery one his owne and proper name.' These proper names agree with the animals' 'seuerall natures' and, like science itself, are not based on chance, but on truth.[58] Henry Arthington echoes Calvin when he calls the ability to name simply *'knowledge'*.[59] The Latin for knowledge is *sciens*, the root of science.

Such a discussion of naming might sound too figurative and sacred to be taken as a central issue in empiricism, but recent work has shown that the influence of the Reformation on the development of empiricism in England is massively significant.[60] Religious ideas are the foundation stone upon which empiricism develops. In *Valerius Terminus* Bacon himself underlines the divine nature of science when he states 'a little natural philosophy inclineth the mind to atheism, but a further proceeding bringeth the mind back to religion.'[61]

Reformed ideas and those of the new science are brought together in a different way in one contemporary sermon; Thomas Adams' *Meditations Vpon Some Part of the Creed*. Here Thomas Adams presents Adam's naming of the beasts in a way which highlights something which is central to Bacon's scientific endeavour. The language the Reformed thinker Adams uses is the language of humanism, but it has implications for our understanding of the new science.

> Thus God gaue the nature to his creatures, *Adam* must giue the name: to shew they were made for him, they shall be what hee will vnto him. If *Adam* had onely called them by the names which God imposed, this had been the praise of his memory: but now to denominate them himselfe, was the approvall of his Iudgement. At the first sight hee perceiued their dispositions, and so named them as God had made them. Hee at first saw all their insides, we

his posterity ever since, with all our experience, can see but their skinnes.[62]

Memory is once again dismissed and the catechitical method revealed as truly postlapsarian. But experience – or sense-empiricism – is also represented as not enough in and of itself.[63] The central element of the new science, sight, is apparently dismissed in favour of the divine. But in questioning the potential of human experience alone Thomas Adams actually fulfils new scientific logic. The perfection of sight which Adam had was not merely human, but divine; and it is to this that Bacon wishes to return.

To call creatures by their true names is not a project of classification (although this was at the heart of the early developments in naturalism in the mid-sixteenth century) it is about supremacy.[64] To name the animals is to truly understand them and for Bacon understanding is where dominion exists: 'human knowledge and human power meet in one.'[65] Thomas Adams represents Adam's naming of the beasts as being inherent in his ability to see deeply. Adam does not merely look at surfaces – 'skinnes' – he looks within, and I do not think that this image is accidental. Adam is the Baconian version of the humanist deep reader. Humanism and science, like theology and science, come together in the concept of naming. Looking beneath the animal skin is the deep reading of the animal, is where humanity proves, in humanist fashion, its human-ness. In Baconian terms, looking beneath the skin of the animal can mean only one thing: experimentation. It is in animal experimentation that naming is achieved: calling creatures by their true names entails entrails.

In Bacon's philosophy the restitution of humanity – the aim of the interpretation of nature – rests upon knowledge of the natural world: on the dissection of animals. Humans exert absolute power in their designation of what is and is not to be experimented upon. This is not a return to perfection as it was in Eden, this is a return to a very different, postlapsarian perfection. Such a perfection in theological terms is a possibility only through Christ. Bacon's science can only achieve its goal of the return to the 'promised bounds' of man's original dominion in the same way that Christ can save humanity: through violence.[66] Either this, or Bacon is saying that the original order of Eden was one of anthropocentrism and viciousness: that biblical perfection itself is not what has been imagined (the term 'imagined', with its implications of fictitiousness, is used deliberately).[67] But the association of Bacon's philosophy with theology – the new science with salvation – is a link which needs further examination because the impact of science on theological ideas is an important one with repercussions for an understanding of human-ness in the period.

IV

Jonathan Sawday makes the link between the new science and theology, but the connection he makes reveals the problem of the significance of animals in Bacon's ideas. Sawday writes:

> Within the framework of a specifically Calvinist theology in six-teenth- and early seventeenth-century England, the anatomists and the theologians were, for a time, able to work with one accord. The scriptures, so they argued, indicated the centrality of the human form to an understanding of God's design. The enquiring researches of the anatomists did not violate this design but sought to trace its detail. The dissection of the human form was not a challenge to theology. When properly understood, it was a sanctified process which was akin to theological reasoning which opened the scriptures to human interpretation.[68]

The idea of the body as a temple of God, a repository of the divine which separates human from animal, re-emerges in scientific writing. Anatomy, Sawday argues, is merely an extension of theol-ogy. Joseph Fletcher called man 'A living *Image*, a quick *Anatomie*'.[69] But the issue of which Sawday does not take account in his link between theology and science is how much was learnt about humans through not only the dissection of human cadavers, but through the dissection and vivisection of animals. While the human cadaver need not compromise the idea of the body as a temple – it is merely an architectural plan – the involvement of the animal body destroys all ideas of separation. The temple of God is sup-ported by the anatomists' endeavour to know but as soon as the animal is placed in the debate the temple, based as it is on a notion of the separation of human and animal, crumbles.

In this way science and religion seem to be at odds: experimentation on live animals in early modern England is a philosophical nightmare.[70] Premised upon the absolute difference of human and animal, on the moral injunction against cutting up the living human which allows the cutting up of the living animal, vivisection is also very clearly based on a sense of sameness: the living animal is cut up to reveal something about humanity. In the mid-sixteenth century Conrad Gesner wrote, it is from animals that 'students in Anatomy … were exercised as in rudiments and grounds, that they might be more prompt, ready, and experienced in the bodies of men.'[71] Bacon himself, in his final, unfinished work *The New Atlantis*, proposed that 'dissections and trials' were carried out on animals 'that thereby we may take light what may be wrought upon the body of man.'[72] Where in humanism the animals in the fables teach the humans morals, so in the experiment the animal sits in for humanity and is, in fact, the original, the human the replacement. It is this paradox of difference and sameness which is at the heart (sometimes literally) of some of the discoveries made by scientists in the early part of the seventeenth century.

The injunction to look, to 'admit nothing but on the faith of eyes' is central to the new science.[73] Following this William Harvey wrote in good new scientific fashion 'I do not profess either to learn or to teach anatomy from books or from the maxims of philosophers, but from dissections and from the fabric of Nature herself.'[74] In a series of just three lectures delivered to the College of Physicians in 1616 Harvey referred to 128 different animals, the knowledge of the anatomy of a large number of which would have come from his own examinations.[75] Harvey's most famous work *De motu cordis – An Anatomical Disputation Concerning the Movement of the Heart and Blood in Living Creatures* – was absolutely reliant on vivisection for its discovery; the movement of the blood could only be observed in a living experimental model. Commenting on the movement of the heart, Harvey wrote:

> Everything is also more evident in the hearts of warmer animals, like dogs and pigs, if you observe attentively until the heart begins to die and to beat more faintly and, as it were, to be deprived of life. Then you may clearly and plainly see the movements of the heart becoming slower and less frequent and its moments of stillness longer; and you may observe and distinguish more conveniently both the kind of movement that it has

and how it is made. In its moment of stillness, as in death, the heart is limp, flabby, and listless, and lies as it were drooping.[76]

The physical progress towards death, the actual stopping of the heart, could only ever be learned at a distance, through the vivisection of animals, and could only ever be discussed figuratively – 'as it were' – at a remove from reality.

But this inability to make the human central in experiments which needed living subjects had other repercussions. Harvey's discovery was about *'Living Creatures'*; man was just one among the dogs, pigs, dolphins, toads, snakes, partridges and eels referred to in the work. The hands-on experience of the scientist was vital to the development of empiricism: without it there could be no enlightening of the 'darkness of antiquity', there could be no sweeping aside of myths.[77] But in sweeping aside these myths the practitioners of the new science spoke figuratively and destroyed the most important myth of all: that of the difference between human and beast.

One of the places where this difference is asserted within Reformed ideas is the human body, the temple of God, and this emerges as a site of sameness in the new science. If the human body is a temple, the animal body – its representative on the vivisector's table – must also be a temple. By logical extension God's image can be found in a dog. The alternative is that the human body is an animal body. As well as this, in distancing human from beast through the deep reading of the animal a form of recognition emerges which undercuts the humanist ideal of deep reading itself. Rather than an assertion of human status this deep reading is a destruction of it. The ability to see the insides as well as the surface brings the human and the animal together in one figure. In total opposition to Sidney's fabled humanity, we are all made animal by our ability to interpret in the new science.

But, as empirical proof of the separation of the human and the animal breaks down in Bacon's ideas another distinction emerges.[78] Using the methods of the new science the scientist displays his human-ness through his ability to name, to know. It is a circular movement: the animal is vivisected because it is different from the human; the human recognises the sameness of the animal through vivisection; another notion of difference – naming – is established to replace the lost sense of physical difference; the name is asserted as the endeavour of the scientific project; the project proposes human potential; and finally, in a return to the beginning of the circle, human potential is illustrated through vivisection, an act which names the animal as animal. So it is the exercise of Baconian reason which distinguishes the human from the animal in Bacon's philosophy. Bacon, in fact, makes humans, in that he is the source of human-ness.

The control of language is a sign of human-ness, and naming – giving the truly perfect and correct designation – is the end product of Bacon's ideas. In *Masculine Birth of Time* the speaker says 'what I purpose is to unite you with *things themselves*'.[79] Where the bestiary must offer a figurative image of the materiality of language – the bear emerges from beginnings and mouths: Ursus, Orsus, ore – the new science has the thing itself. Science returns humans to the status of gods. As Marion Trousdale notes, 'God's word is substance, it is ontologically real. But in man language is accident, not substance.'[80] In the new science the object and the word coincide, there is no distinction between the signifier and the signified, the inside and the outside, or between the animal and the human. What exists beyond, as a third term – the referent – is the experimenter. There will always be someone performing the experiment, seeing the insides and the outsides, and it is the experimenter who is Bacon's human: 'for whensoever he shall be able to call the creatures by their true names he shall again command them'.[81] Adam, the original namer, is the great vivisector: he sees beneath the skins. Bacon's Eden is never Edenic for the animal.

In *The New Atlantis* Adam's role is fulfilled by the Father of 'Salomon's House'. He rises above the animal in his total knowledge of Bacon's ideal college. There are four institutions on the island in this text: the college, the church, the family and the state itself. The latter three are brought together in two of the men encountered in the narrative.[82] The governor of the state's House of Strangers – a Christian priest by vocation – is 'parent-like' in his care of Bacon's narrator and his fellows.[83] The Tirsan (the father of his family) is priestly in his blessing and statesman-like in his solving of family squabbles.[84] But the Father of Salomon's House is the parent, the priest, the statesman and the scientist all rolled into one. In Bacon's perfect commonwealth science is not a development of religion, it is the institution which encompasses every other institution, including the church. Science is the thing itself. Thomas Adams' dismissal of human experience is explained because Baconian experience is more than merely seeing. It is seeing everything: it is the difference between a glance and a thorough examination. Julian Martin argues that Bacon's 'motives for reforming natural philosophy and for reforming English law were the same: the desire to create the conditions for empire'.[85] The link between Bacon's statecraft and his science is undeniably important, as Martin shows, but in emphasising the legal foundations of the new science Martin tends to place the conquest of nature in the background. It should be remembered that in *Novum Organum* the extension of empire is regarded as 'covetous' while the extension of the power of the whole of the human race is 'without doubt both a more wholesome thing and a more noble'.[86] Naming the creatures is the aim of Bacon's ideas: animals and not land are the first colony of the new science.

But Bacon is also presenting empiricism as a religion. As with Christianty the evangelism of the scientist is vital – the narrator is sent out into the world to spread the word about the new science in *The New Atlantis*. He is given, as it were, a Bible. The word of Bacon, like the word of the scriptures, must move beyond the confines of the place of worship. Bartholomew's dissemination of the word of God on the New Atlantis destroyed the linguistic chaos (Babel), likewise Bacon's ideas return humans to their original state of substantial language.[87] As in Reformed Christianity, where some are reprobate, Bacon does not propose human status for all. Within Reformed ideas only some are saved, within Bacon's philosophy only some are experimenters.

But *The New Atlantis* is not merely a document of faith, it is also a fiction, a fable, as William Rawley termed it.[88] This narrative of shipwreck and recovery is a tale told by a traveller: 'We sailed from Peru ...'.[89] Its fictitiousness is vital: one of the foundations of the new science – as it appeared in its early form in Gesner's work – is the anecdotal recollection of the wayfarer, the stranger. What the *New Atlantis* seems to ask is whether the anecdote of the way-farer is fiction or fact.[90] As modern criticism asks whether John Mandeville ever existed, and if he did, whether he ever left his library, so we can ask whether *The Travels of Sir John Mandeville* and *The New Atlantis* are actually very different.[91] Empirical truth is celebrated on this fabulous island even as its basis is brought under question. It takes a fable to displace the fables of the old science. In fact, Bacon's human emerges in a fable as a fabler. The experimenter, paradoxically, mirrors Aesop.

V

The title page of the 1570 edition of *The Fables of Esope in English* expands on the short title of the work with a further explanation of its contents. The long title shows that the work also contains:

> all his lyfe and Fortune, howe he was subtyll wyse, & borne in Grece not farre from Troy the greate, in a towne named Amoneo, he was of al other menne moste defourmed and euill shapen. For he had a great head, a large visage, long iawes, sharp eye[n], a short necke, crokebacked, greatebelly, great legges, large feete. And yet that which was worse, he was dombe and could not speak. But notwithstandinge this he had a singuler witte and was greatly ingenious and subtill in cauillacions, and pleasaunt in woordes, after he came to his speache.[92]

The Life of Esope, as this part of the text is known, first appeared in the fourth century B.C., two centuries after the attribution of the collection of fables to Aesop. In the second century A.D. 'a long and elaborate *Life* was written [which] ... survives in a tenth-century manuscript.'[93] The *Life* makes the author of the beast fables, the man who made the beasts speak, at birth, an animal. As Louis Marin has noted, he 'does not have – or barely has – a human form because he does not speak'.[94] We are back to the humanist alignment of eloquence and human identity. Bacon's theory of deep reading can also be traced back to humanism, and, similarly, the narrative of Aesop coming to language offers a parable which is a parallel of Bacon's version of the human potential to reclaim what is lost. The humanist and the empiricist both offer the same tale of empowerment through language, but one of them does it in order to tell stories, the other to banish them.

The story of Aesop's early life is from silence to speech and from slavery to freedom: speaking means liberation. Just as Orson is able to communicate through signs before he gains spoken language, so Aesop can also communicate before speech, but, again, like the wildman, this communication is not free, it is tied to the body: '[w]ords are not substituted for things ... but things are substituted for words'.[95] The *Life* begins with the story of Aesop and the figs. Agatopus and his fellow, two other slaves of the household in which Aesop works, eat the figs which their master has asked to be put aside for later: they are the forbidden fruit. The crime of eating the figs is, they think, committed with impunity; 'we shall saye to [the master] that Esope hath eaten the[m]. And bicause he can not speake he shal not excuse him selfe, & therefore he shalbe wel beaten'.[96] But Aesop does manage to excuse himself even without speech – just as Orson could communicate to say that he could not communicate. He drinks warm water and throws-up the contents of his stomach and then challenges Agatopus and his fellow to do the same. The material objects – the actual, rather than spoken, contents of the stomach – are revealed to prove Aesop's innocence, while, against their will, Agatopus and his fellow also empty their stomachs and prove their guilt. The materiality of communication – the use of things and not words – distances Aesop from his fellows, but it also allows him access to the truth.

After a Good Samaritan-like meeting with a priest, however, Aesop is rewarded for his generosity by the Goddess of Hospitality. He is given 'sapience and habilitie: And also ... the gift of speach'.[97] On awaking from his divinely inspired sleep Aesop says:

> I haue not onely slepte, nor swetely rested but also I haue had a fayre dreame, and without any impechement, I speake, and all that I see I call by theyr proper names as an horse, an oxe, an asse, a chariote, and to al other things I ca[n] to euerich geue his name.[98]

This moment, Annabel Patterson writes, reveals 'the theory of language built into the ancient *Life* of the Father of the fable'. She argues that 'the fabulist mysteriously recovers the Adamic prerogative of differentiating (naming) the creatures'.[99] This reading of Aesop's naming fits with Bacon's notion of humanity. The Father of the fable and the Father of empiricism coincide: where Aesop is freed from his silence when he gains speech from the Goddess of Hospitality, Bacon's humans are liberated from the confines of the Fall through the re-assertion of the naming process which can likewise only be achieved through the inspiration of divinity (Christ or Bacon).

But the narrative of the figs questions Patterson's linking of Aesop's coming to language with the Adamic prerogative. In the story of the figs the veracity of Aesop's communication was

absolutely embedded in his *inability* to speak. Marin, who calls
Aesop the Aesop-animal, writes

> The animal can provide no counter-narrative, no counter-expla-
> nation. However, let us imagine that possibility – let us give
> Aesop speech for a moment. What could Aesop have *said*? 'I
> did not eat the figs ...' .... 'Prove it,' the master would have said.
> It is impossible for him to produce proof: the *corpus delicti* has
> disappeared.[100]

But Aesop, of course, does produce proof – he vomits it up. The
materiality of the proof is its proof: spoken language can never offer
such unassailable truth. The *Life* reveals that the move is not from
silence to substance, but actually from substantive (material) communication – vomit – to naming through the fluidity of spoken language.[101] The concrete reality of communication is abandoned with
the receipt of speech and in the *Life* the animal gains a voice in two
senses. Literally, Aesop the animal speaks; figuratively, Aesop,
having lost material language, writes the truth through fables in
which the beasts converse.

This difference from the biblical narrative of origin is emphasised
later in the *Life* in the tale of the tongues. In this tale Aesop plays a
trick on his owner Erantus which proves his wittiness in spite of his
slavery and physical deformity. Aesop is sent out to buy 'the best
meate that [he can] fynde' for a party his master will throw for 'the
scolers'. He purchases 'the tonges of swyne and of Oxe[n], and
dight the[m] with vineger, & set the[m] on the table'. When these
are eaten he brings out more tongues which are dressed differently,
much to Erantus' chagrin. Aesop explains his actions:

> I wold faine know of the philosopher what is better tha[n] a
> tongue. For certainly, all arte, all doctrine, & all philosophie be
> notified by the tongue, without the which ther could be no ioy
> nor co[m]pany amo[n]g men, for by it the lawes are declared, but
> if the good receyueth praise, the euill rebukes, the sorowful
> co[m]fort, the folish instructio[n], the wise men knowledge. And
> finally, the greatest part of ye lif of mortall men is in the tongue,
> & thus there is nothing better than the tongue, nor nothing more
> swete no better of sauour, no more profitable to me[n].[102]

Aesop is forgiven, and the lesson of this embodied fable is learned.
The next night Erantus throws another dinner party to make up for

the poor food at the first one. Again he asks Aesop to buy meat, but this time asks for 'the worst meat that thou canst finde': he thinks he can read Aesop. But Aesop again buys tongue and his excuse this time reveals the dangers of language. Aesop asks 'what is worse or more venimus tha[n] an euil tong. By the tonge me[n] be perished, by ye tongue they come into pouerty. By the to[n]gue cities be destroied. By the tonge cometh much harme.'[103] The spoken word produces company among men but it is also anti-social. The tongue is truly 'ambivalent'.[104]

This ambivalence finds a point of origin in both the *Life* and the Fall, and the results of these two sources link the Aesop-animal with Bacon's human. For Aesop the loss of substantive communication is found in fables: it is the animals who end up speaking truths, giving voice to substantive statements, with all the ironies which that holds for humanism. Bacon proposes that the return to the ideal of calling creatures by their true names will restore human dominion. Animal experimentation is one way in which this postlapsarian process takes place. But in cutting up the animal, in attempting to return to substantive language, what is revealed is the animality of humans. The experimenter names animals, but he also names himself. Bacon's human ends up, not exercising dominion over animals, but speaking, like the Aesop-animal, with the voice of the animal who he should be able (in terms of the Adamic prerogative) to name.

Bacon's final celebration of Baconian science and the Baconian scientist takes place, as noted, in an unfinished fable told by a fictional traveller. In *The New Atlantis* the insubstantial nature of Bacon's claims becomes clear. At the sight of the miraculous 'pillar of light' which appears to bring the islanders to the word of God one of the wise men of Salomon's House is inspired and praises God that they are able 'to discern (as far as appertaineth to the generations of men) between divine miracles, works of nature, works of art, and impostures and illusions of all sorts.' Divine inspiration reveals the difference between fact and fiction. The column of light before him, he says, is God's 'Finger and a true Miracle'.[105] This miraculous pillar, like the star at Christ's birth, points humanity to the scripture and to salvation. *The New Atlantis* itself is just such a miraculous column of light. It is an imposture which announces the path to truth or is, if you like, Bacon's 'Finger and a true Miracle'. Its fabulous nature undermines the notion of truth which it proclaims: the difference between illusion and nature, mirage and reality, human and animal is sought through fantasy.

This is a problem for the construction of the human, because if the enclosed truth of Bacon's philosophy is broken – if Bacon is unable to declare Baconianism to be proof of human status – and if scientific truth – the ability to see insides and outsides at once, to call creatures by their true names – reduces humans, as William Harvey noted, to the status of a creature, then all notions of difference have gone. Bacon, while pronouncing the human to be the anti-fabler, like Aesop is merely a fabler, and the characters he uses to tell his tales are, like the Aesopic characters, merely animals.

Where Bacon would appear to be asserting an uncompromising return to power, a separation of human and animal, he is, in fact, unable to do such a thing without invoking the very thing he is leaving behind. His human is revealed as a physical animal on the experimenter's table, and when he speaks he has the fantasy of substance, but always discovers the substance to be in reality 'as unstable as water and as gusty as wind'.[106] While advocating the return to the original state of dominion Bacon's ideas actually reveal that human separation from the animal is a myth, that neither the body nor the mind can be offered as the site of human-ness because neither is provable within Bacon's scheme in which proof – the matter of fact – is central.[107]

When the human is separated from the animal – when it is decided what is and is not to be placed on the experimenter's table – what are invoked are not 'neutral, unbiased and objective' truths about difference, but fables, stories humans make to tell themselves they are central.[108] Once again anthropocentrism results in the very thing which it seems to deny. In Bacon's anthropocentrism it is not so much the animals who are made human, rather it is the humans who are revealed to be animal.[109] To read beneath the surface – the humanist ideal – is to strip away a layer of human-ness. The body, the thing which Reformed thought claimed differentiated the fallen human from the animal, is revealed to have no such role in Bacon's philosophy. The body returns, however, in a different area of English society. In Edward Coke's legal definition of the inheritor – the person with the right to possess property and to exercise dominion over worldly goods – it is presented as the differentiating feature. The logic behind this assertion is the subject of the next chapter.

5

The Shape of a Man: Knowing Animals and the Law

At the end of the sixteenth century Pierre Le Loyer recalled a legal case in which a monster's right to inherit came under scrutiny. The case centred on a child who was deformed because he was conceived, Le Loyer writes, when his father was still wearing the costume of 'a divell' he had worn for a play. His monstrosity was the result of 'some vehement imagination ... or else ... a very feare which seized vppon' his mother.[1] Le Loyer describes the son as 'so monstrous, as in his countenance, his head, his face, and all the other parts of his body, especially his feete, hee resembled and was more like vnto a Satyre ... then vnto an ordinarie and naturall man'.[2] In the court case the monster's younger siblings attempted to exclude him 'not onely from the birth-right of being heire and eldest sonne, but even from the totall succession of any thing that hee should claime'.[3] The defence which the monster offered, and which was upheld by the court, proposed that 'albeit he had his visage and some partes of his body in some sort deformed: yet that was no reason that hee should bee helde and reputed as a monster, seeing hee had the vse of reason and humane discourse'.[4] The monster claims human status by asserting his mental capacity and by discounting his monstrous body. The removal of the emphasis on the physical being and the assertion of the primacy of the two things which humanism had claimed as the qualities which distinguish human from animal – 'the vse of reason and humane discourse' – win through in this case. The monster is able to inherit, is recognised, in fact, as human because the mind is understood as the seat of human-ness.

The monstrous brother, however, adds another argument to his defence which reveals the importance of his right to inherit:

> Yet is it not to be yeelded in any sort, that those monsters that are borne of any man, should be slaine, as the Appellants doe seeme to affirme; whether they have the vse of reason ... or whether they have not the use of reason, but be so monstrous, as they have not so much as the face of a man, but rather of some beast ...[5]

His siblings' suggestion of the necessary death of all monsters reveals a potentially fatal connection between not inheriting and monstrosity. In asserting his human status and right to inherit, the monster asserts his right to live.

This link between the right to inherit and human-ness is one which can also be traced in the writings of Edward Coke. To inherit is to gain property, and possessors of property have a clear status within the law. The law serves, in many ways, to protect their interests by asserting an important difference between owner and not-owner (thief, trespasser). The distinction offered here is between humans. When we look to differentiate humans from animals it is more straightforward: the categories are owner and owned. The human has a right to possession which the animal does not, and humans can possess animals. It would seem simple to use this distinction to delineate the human in the law, but Coke does not do so. With no obvious link to Reformed ideas about the temple of God he proposes that the body, not the mind or the status of ownership, is the distinguishing quality of the human. He argues that the physique is proof of the ability to inherit which in turn is proof of human-ness.

> A monster which hath not the shape of man kinde, cannot be heire or inherit any land, albeit it be brought forth within marriage, but although he hath deformitie in any part of his bodie, yet if he hath humane shape he may be heire.[6]

For Coke, the possession of a human shape entitles subjects to possess and maintain property: gives them power over everything which is not human. The similarity of human and animal bodies which emerges in the deep reading of the animal in Bacon's philosophy is replaced by an examination of the exterior, rather than interior, physique. Coke refuses to look beneath the skin.

The egalitarian implications in his shift in emphasis also remove Coke from another model of human-ness which existed at the time:

> One that is borne deafe and dumbe may be heire to another ... And so if borne deafe, dumbe and blinde ... . Ideots, leapers, madmen, outlawes in debt, trespasses, or the like; persons excommunicated, men attainted in a *praeminure*, or conuicted of heresie, may be heires.[7]

Not only are the dumb, those incapable of human discourse, included within Coke's delineation of the human; even idiots and heretics have the right to own property. Implicitly those without reasonable minds have human status if they have human shape. This is a sentiment which goes against the findings of the case La Loyer reports, and against other constructions of human-ness which we have so far encountered. In Coke's distinction the surface – shape – is the site of stability.

This chapter examines Coke's emphasis on the body in his definition of the property owner and argues that the reason for such a delineation of the human is to be found in laws dealing with the possession and status of animals. Owning an animal would seem to be an exercise of dominion without question: the animal becomes merely an object, like a table or a chair. A close examination of the status of animals as property in the law reveals, however, that to own an animal the owner has to know the animal, and in knowing the animal the owner not only exercises his mental power he also gives the animal a character. Asserting that the animal is a possession does not objectify it, rather it entails recognising the individuality and mental capacity of the animal. This recognition narrows the gap between owner and owned and destroys the function of these categories to make human difference and superiority.

The power of the human mind and the link with the beast also emerges in a very different but related way in legal discussions of the status of the idiot. In these debates the legal distinction between accidental and intended actions places the idiot alongside the animal as incapable of willingly committing crime. But, while the idiot is regarded, like the animal, as inherently not a criminal, unlike the beast, he is still able to possess property. In this nightmare representation of the human where notions of both ownership and criminality fail to define the species all that is left to distinguish human from animal is that most fragile of signifiers, the body. The animal in the law does not support a difference: it destroys it. Before the issue of the legal status of the beast can be examined it is worth looking at the theories surrounding the systematising of the law in early modern English culture, because in them other fears about the capacity of humanity emerge.

I

In 1615 Sir John Davies noted that the common law 'hath ever been preserved in the memory of men, though no man's memory can reach the original thereof'.[8] Up until the early seventeenth century the common law had been 'an improvisation in the face of political changes'.[9] Notions of inherent truth, of inherent right and wrong, broke down in the face of these misrememberings and improvisations. Coke wrote that 'Nothing is or can be so fixed in mind, or fastened in memorie, but in short time is or may be loosened out of the one, and by little and little quite lost out of the other'.[10] There was a need for the creation of a coherent system, for the increase of certainty. The law needed to be written down in order that it could be commonly known.[11]

The separation of law from mind – the fact that it is not regarded as an innate part of the understanding – is exemplified in Coke's theory of artificial reason. Coke is alleged to have told James I that the subjects' fortunes 'are not to be decided by natural reason, but by the artificial reason and judgement of law, which law is an act which requires long study and experience before that a man can attain some cognisance of it'.[12] This is at once a reminder of Coke's belief in lex before rex, law before king – the source of his disagreement with Bacon – but it also suggests that natural (innate) reason counts for nothing in the law. There are, John Underwood Lewis argues, two ways of viewing legal reason: for Thomas Aquinas 'the law is an "ordinance of reason," a directive judgement that guides men in their choice of means to be used in the attainment of social goals'. For Coke, however, the notion of reason in the law refers to its '*reasonableness* rather than ... to its being essentially a rule or principle of human action.'[13] The law exists as a source of reason external to the human rather than as a reflection and systemisation of innately human reason. For Coke human actions are reasonable only in so far as they are recognised as such by the law. There is no reason without the law. So, where there might be some agreement between Coke and humanist educationists over the belief that true understanding comes through labour, and between Coke and Bacon over the importance of experience, Coke departed from the belief in human potential which both propose when he argued that humans' natural judgement is incapable of making legal judgements.[14] Artificial reason, something external to the individual, was needed for justice to exist.

Such an interpretation allows Coke to propose the common law to be ancient and inviolable. It is a body of ideas which preceded him and will outlive him. The law is not bound by the temporal claims of individual monarchs or individual citizens: once found, legal truth is a scientific and constant truth. Coke's *Reports* (eleven parts of which were published between 1600 and 1615) merely show how firmly established the truth is. They use the inductive method, familiar from Bacon's philosophy, to establish general theories through an examination of the specific case.[15] But in calling these scientific truths artificial reason and proposing the inviolability of the law, Coke emphasises the inability of the human mind: the law, despite his assertion that it is ancient, can be forgotten.

The written word was central to the establishment of the law. At its most simple the ability to read could save the convicted felon from execution through benefit of clergy – the ability to 'read' the first lines of Psalm 51 ('neck verse').[16] But in what Coke called 'a writing or a scribling world' the written word also recorded that which would be lost by the human memory: 'It is therefore necessary that memorable things should be committed to writing (the witnesse of times, the light and life of truth) and not wholly be taken to slipperie memorie which seldome yeeldeth a certaine reckoning.'[17] Slippery memory, like the slippery tongue of the previous chapter, reveals the fragility of the human mind, and this fragility finds a parallel in Coke's emphasis on the body in inheritance laws. If the law itself exists outside of the mind because the mind is so naturally incapable of making judgement then the status of the human must also be sought outside of the mind, in the body.

But this depiction of human-ness existing in the body was not without its problems. It was not only the memory which was frail; if ancient writers had not written the law 'certainely as their bodies in the bowels of the earth are long agoe consumed, so had their graue Opinions, Censures, and Iudgements ... long sithence wasted and worne away with the worme of oblivion'.[18] The body, the seat of human-ness in Coke's work, like the unwritten law (*jus non scriptum*), is shown to be transitory and corruptible. The thing which defines the human also reveals the status of the human to be temporary, slipping as the laws slips in the memory. When we die the symbol of our status, the body, dies with us. When the law is forgotten human-ness – which the law defines – is lost as well. The importance of the physical imagery Coke uses can also be traced in his definition of truth, and here, just as in his depiction of the unwritten law, Coke represents truth as something physical and corruptible: 'for her place being betweene the heart & the head doth participate in them both, of the head for judgement, and of the heart for simplicitie.'[19] In utilising this physical imagery Coke proposes that truth, like the law, can decompose in the grave with its maker, but it can also, if written, outlive the body in the form of the report. Being written, in fact, is everything. When the law is unwritten humans are disembodied and true possession is impossible.

Thomas North argued that fables were 'an artificiall memorie', a way of learning judgement through the embodying of truth in the animal fable.[20] By depicting reason rather than memory as artificial, however, Coke once again goes beyond the logic of humanism. The memory is a part of the understanding, reason is the understanding itself. In Coke's idea of the law as artificial reason we have abandoned the notion of the inherent potential of the human mind altogether. Recognising the impossibility of a coherent legal system emerging from the understanding working alone, Coke set about making a new and artificial reason which was external to the human. The *Reports* record the actions and judgements of the law. They enact remembrance and monumentalise the transitory. In fact, a legal report writes the truth.

This raises the question of whose truth was made in the law.[21] When Coke was putting together his *Reports* from contemporary cases and from the work of predecessors such as Plowden, Dyer, Brooke and Bedlowes they were very clearly labelled *Les Reports De Edward Coke*.[22] In this sense the truth, that apparently absolute and unquestionable thing, is revealed as subjective, is made by an individual. But Coke writes that he has published the work for those who heard and knew the cases; for those who heard but had forgotten the cases; 'and lastly for the common good (for that is my chiefe purpose)'.[23] Just as the law itself is both ancient and forgettable, so Coke's work is an act of (anonymous) social duty and an act of authorship.

But the subjective nature of the truth in the *Reports* does not actually undermine its importance. The subject knew what was reasonable because it did or did not conform to legal principle: an external force defined what was considered to be an internal act. There is an acceptance that truth is both ancient and, as Bacon noted in *Novum Organum*, man-made. But if the law, like truth, is in part man-made then so are its judgements, and it is those judgements which are making humans. The human of the early modern common law is not merely to be defined in terms of the physical body, it is also to be recognised as making and defining itself. There is no sense of the God-given difference between the species in the law, there is only a difference which is made by humans themselves, and as humans can make their own status, so they can also destroy it. These man-made problems surrounding the status of humans emerge most fully in the laws dealing with the ownership of animals. Before getting to the issue of ownership, however, in good Baconian fashion I need to dismiss some myths about early modern legal conceptions of animals.

II

The trials of animals which took place in continental Europe throughout the early modern period seem to present the animal as a sentient creature, capable of understanding the law.[24] They give, as one historian understates it, 'an impression of anthropomorphism'.[25] The animal was held in a human prison, was given a lawyer, and was – after the verdict – executed on a human scaffold. What these trials seem to imply is that the animal showed intention, revealed malice and an understanding of the nature of the act it had allegedly performed, and that it therefore warranted full legal treatment. The situation 'invites us to reconsider the boundaries which are conventionally drawn between humans and animals'.[26] Another understatement.

It would be very easy to dismiss the anthropomorphism of the trials of animals as evidence of the ignorance of the past and the civility of the present. But the suggestion that people could not differentiate a pig from a human until the emergence of modernity is obviously ridiculous. In fact animals were tried, not because they were considered culpable, but because they had revealed a fragility within a very important human institution. The law was in place to create and control behaviour: it was 'the ideological cement which held society together'.[27] If an animal, albeit unwittingly, committed a criminal act the law must be seen to punish. The stability of the law itself lies at the centre of the trials of animals.

In prelapsarian Eden there was only one law, the sanction against eating the forbidden fruit. After the Fall, however, new laws were instituted, and these involved, of course, human relations with the animals. Exodus 21: 28 reads 'If an ox gore a man or a woman, that they die: then the ox shall be surely stoned, and his flesh shall not be eaten; but the owner of the ox shall be quit.' As J. J. Finkelstein has shown, this scriptural injunction was to emphasise that

> Simply by its behavior – and it is vital here to stress that intention is immaterial; the guilt is objective – the ox has, albeit involuntarily, performed an act whose effect amounts to 'treason'. It has acted against man, its superior in the hierarchy of Creation, as man acts against God when violating the Sabbath or when practising idolatry.[28]

What is also important is the quittance of the owner. The responsibility for the crime is not the animal's, nor is it the owner's, because responsibility is not the issue. The issue is maintaining the order of the human community. Animals were not free from social rules (that would be dangerous) but were 'subject to the universal law which placed them below man'.[29] Anthropocentrism, again, is central.

But this does not really explain the need for a full-blown trial of the animal. For this we can return to E. P. Thompson's ideas about power (referred to in the introduction) where he proposed that what is most at risk in society is most staged. Reinforcement is only necessary if what is being reinforced is seen to be fragile.[30] Just as executions 'underline the power of the state', so the animal trial reveals both the fragility of human supremacy over the beast and enacts the power of humans.[31] A felonious animal had violated the dominion of man and a form of social cleansing was necessary to reassert 'peace' in the human community.[32]

The execution of the animal was a spectacular affair; as in monkey-baiting, the animal was sometimes dressed in human clothes.[33] But, again like monkey-baiting, this potential confusion over species was averted because the difference between anthropoid and *anthropos* was still in place. The dressed pig was always a pig in a dress, but it was a pig with legal rights. The apparent anthropomorphism of the animal trial argues for a common link between human and animal defendants, but ultimately serves a very specific role. The admonition was directed at humans: 'plenty of *people* saw what happened, but ... no one ever brought another pig to see the execution'.[34] Even an animal was subject to the law. The trial and execution were a reiteration of human power. No animal was ever found not guilty: animals were 'never "tried" in the strict sense', the punishment which followed was 'mandated by the simple fact that the [crime] had taken place'.[35] The animal trials represent not a sense of animal volition but a symbolic restitution of human order.

However, these trials took place in continental Europe. English law had a different way of dealing with animal 'malefactors'. In England the deodand law (from the Latin *deo dandum*, given to God) was called into action 'when any moveable thing inanimate, or beast animate, doe move to, or cause the untimely death of any reasonable creature by mischance ... without the will, offence, or fault of himself, or of any person'.[36] The animal is placed alongside inanimate objects and is differentiated from the 'reasonable' creatures which it might injure. Any death dealt with under the law of deodand is represented as 'Casuall death', as opposed to murder or death by misadventure.[37] There is no sense of intent, but there is a trial, a 'lawfull inquisition of twelve men', and if guilt is asserted then 'being *precium sanguinis*, the price of blood, [they] are forfeited to God, that is to the King, Gods Lieutenant on earth, to be distributed in works of charity for the appeasing of Gods wrath.'[38]

A case of deodand which is reported in the Essex records during the reign of Elizabeth I concerns some inflated bladders which failed to keep a young boy afloat. The boy and two of his friends were drowned and 'the bladders worth 1d are deodand'.[39] In another case from 1619 it is live animals – not their inflated body parts – which are at the centre of a case: 'Two men riding over the river of *Trent*, were drowned by the violence of the water; It was moved for the King that their horses should be *Deodands,* and denied *per totam curiam* [by the whole court].'[40] The water and not the horses is adjudged responsible and the appeal turned down. Where, in continental Europe, an animal on trial was always guilty, in deodand legislation an animal could be declared innocent.[41]

But, in fact, it is not the animal who is declared innocent; that would imply that the inflated bladders could also be declared innocent. It is, rather, the owner of the animal who is innocent. Deodand legislation never asks questions about the animal's intention, what is at stake is the status of the owner of the animal. The deodand demands a responsibility on the part of the human, and if that responsibility cannot be fulfilled then the animal must be given up; 'the liability only lasted so long as the owner retained his ownership of the offending beast'.[42] The animal is the possession which reveals the owner's capacity as a human. The animal does not know the law – it is an animal – and a human can prove his human-ness by recognising the law and his responsibility within it. The law proves the reasonableness of the owner. Should the human fail to take this responsibility which his status brings with it, then that status – as property owner, possessor of dominion over wordly goods – is lost. As in Reformed ideas of double justification, there are two stages of human-ness evident in deodand legislation. There is the humanising status of ownership itself, and then there is the responsibility which ownership brings. The first is primary, but both are necessary in the achievement of human status. Failing to take responsibility for the animal is a failure to be human.

The importance of the owner's knowledge of their animal, their ability to take responsibility, emerges again in the law dealing with the action of 'scienter', or knowing (the Latin root of the term science), which relates to 'a state of affairs which the defendant knew or ought to have known was dangerous'.[43] Sir George Croke records a case in which the 'Defendant kept a Mastiffe, sciens [knowing] that he was assuetus ad mordendum Porces [likely to bite pigs]'. The plaintiff's sow died after being bitten and the case was found for the plaintiff with the judgement that 'it is not lawfull to keep Dogs to bite and kill Swine'.[44] The owner of the dog should have known his animal, and it is he who must pay the penalty for acting without responsibility. The dog is given up: it is property which the owner has forfeited the right to possess.

The English law, then, moves in two distinct directions in its representation of animals. The emphasis on the status of the animal as property appears to make it much less anthropomorphic than continental European law: the animal itself is never on trial. But where the continental European trials of animals are not to be interpreted as asserting animal culpability, in England the emphasis on ownership and the status of the animal as property brings new and, paradoxically, anthropomorphic problems. The animals become individuals whom the owners have to know. The animal, in short, is given a character.

III

In the early modern English law there are three categories of animal: *ferae naturae*, wild animals; *domitae naturae*, domestic animals; and recreational animals, whose lack of a Latin name represents their lack of status in the face of the law (this is examined further below). Throughout the period these three categories are central to the legal status of animals and to the possibility of owning them.

In 1618 Michael Dalton offered a definition of the different categories and their meanings, and I will use this as an introduction to the problems faced in asserting property rights over animals. There are three classes of ownership within the English law; absolute, qualified and possessory. The first, absolute, is clear: this class of ownership 'a man cannot haue in any thing which is *feræ Natura*, but onely in such things as are *domitæ Natura*'.[45] Working animals may be possessed absolutely. They are equivalent, William Lambard points out, to 'housholde stuffe'.[46] In the second and third classes of ownership – qualified and possessory – things are less clear: I quote Dalton at length.

> [Qualified or possessory ownership] a man may haue in things *feræ Natura*; and to such properties a man may attaine by two meanes ...
> 1. By industrie: and this may be eyther by taking them onely; or by making them tame ... But in these last a man hath but a qualified propertie ... so long as they remaine tame, and so long felonie may be committed by taking of them away; but if they attaine to their naturall libertie, and haue not *animum reuertendi* [the status of things reclaimed], then the propertie of them is lost.
> 2. *Ratione impotentiæ et loci* [with regard to weakness and location]; As where a man hath young Goshawkes, or Herons, or the like, which are *feræ Natura*, and do breed (or ayre) in my ground, I haue a Possessorie propertie in them; for as if one takes them when they cannot flie, the owner of the soile may haue an action of trespasse ...[47]

Owning a wild animal asks for the engagement of the human in a way which the ownership of, say, a cow does not. But a wild animal is still an object to be possessed. In order to possess a wild animal that animal has to be 'reclaimed'; it has to be tamed and owned, and the use of the term 'reclaimed' to describe this relation reveals the importance of the Fall in the possession of the natural world. To own an animal is to claim back the prelapsarian relation with that animal. Ownership is, as Bacon said of naming, 'a restitution and reinvesting (in great part) of man to the sovereignty and power ... which he had in his first state of creation'.[48] In the law notions of tameness do not mean that any kind of behaviour-test is carried out; in most cases physical marks of ownership are enough to prove that the animal is reclaimed. Reclamation is more about human control – marking – than animal nature. The reclaimed paradise of the English law, like that of the new science, is hardly a return to paradise for animals.

However, the laws dealing with marking were not as cut and dried as might appear. Marking was sometimes deemed to be unnecessary, as two appeals brought before the King's Bench reveal. In the first, 'Sir Francis Vincent *versus* Lesney' (1626), Lesney has been fined six pounds for hitting Vincent's hawk and killing it. The appeal does not question that the death of the hawk followed Lesney's actions, but argues that the prosecution did not show in the original case that the bird was reclaimed and that without such proof there should be no damages paid; an unreclaimed wild animal is not regarded as any class of property in the law. This appeal was turned down by the judges with the simple explanation that Vincent has property in the hawk 'which he may only have who hath the possession'. Vincent had obviously fulfilled Dalton's third class of ownership.[49]

In another appeal dealing with the theft of a hawk the issue of the status of the animal is raised again when the defence appeals on the grounds that the plaintiff 'doth not shew, That she was a reclaimed Hawk, and made tame, nor that she had Bells or Vervells to shew who was her owner'. The defendant calls for proof that the hawk which he has 'converted to his own use' (not a crime in itself if the bird was not reclaimed by another at the time of the conversion) was the property of the plaintiff and therefore already reclaimed. This is proof which can be maintained by physical signs of ownership – bells and vervells (ankle bracelets). Once again, however, the appeal is turned down. The King's Bench argues that the defendant knew the bird belonged to the plaintiff with or without physical signs of ownership being called into account.[50]

Coke writes that in the case of a lost hawk vervells might prove 'reasonably' that it is reclaimed. He goes on to say, however, that such marks are not compulsory: 'Albeit these Hawks, that shall be so lost, have no Vervels, yet must the finder carry them to the Sherif, for Vervels are not required by this Act.'[51] This places more emphasis on the actual hawk itself – the fact that it is a hawk – which would imply that all hawks are owned, and would totally undermine the distinctions which place the hawk in the category of *ferae naturae*. Coke recognises the difficulties which underline such discussions of marks of ownership. There is a legal contradiction, and he states that ultimately the property status of hawks (and other wild animals) is an impermanent one. Even if they are 'reclaimed for delight and pleasure ... they may become wild again, and return to their naturall liberty'.[52] Ownership of *ferae naturae* is like postlapsarian language: it never achieves true fulfilment. It is a reminder of the fact that reversing the implications of the Fall – allowing, in this instance, humanity to reassert dominion over animals through ownership – cannot be fulfilled merely through the exercise of human control and adornment. Possession, although in some cases nine-tenths of the law, is fragile. The wild animal cannot be wholly tamed, and, because of the intimate relationship which is established between tameness and property, a wild animal can never be wholly owned. The status of the human (the possessor) breaks down in the face of untamed nature.

Tamed nature, *domitae naturae*, should offer a different and more comfortable position for humans. Cattle – the generic term for domestic animals, related to chattels, a French term with the same root as the word 'capital'[53] – were for human use: there were no wild sheep in England. As Leonard Mascall wrote in 'A praise of Sheepe':

> These cattel (sheepe) among the rest,
> Is counted for man one of the best.
> No harmeful beast nor hurt at all,
> His fleece of wooll doth cloth vs all:
> Which keepes vs from the extreame colde:
> His flesh doth feed both yong and olde.
> His tallow makes the candles white,
> To burne and serue vs day and night.

His skinne doth pleasure diuers wayes,
To write, to weare at all assayes.
His guts, thereof we make wheele strings,
They vse his bones to other things.
His hornes some shepeheardes wil not loose,
Because therewith they patch their shooes.
His dung is chiefe I vnderstand,
To helpe and dung the plowmans land.
Therefore the sheep among the rest,
He is for man a worthy beast.[54]

The poem ends with 'FINIS', which sums up the state of the sheep itself. The animal is totally useful to humanity; nothing is wasted. This hang-over from the old science is reflected in the law in which the ownership of domestic animals is apparently straightforward.[55] In 1592 William Lambard wrote:

Money, plate, apparell, housholde-stuffe, corne of any sort (or haie, or fruit) that is severed from the ground, horses, mares, coltes, oxen, kine, sheepe, lambes, swine, pigges, hens, geese, ducks, peacockes, turkies, and other beasts, and birds of domesti-call (or tame) nature, are such, as felonie may be committed in the taking of them.[56]

Animals are valued for their use; are aligned with domestic objects such as furniture; and are clearly regarded as non-sentient. This, however, is complicated when the necessity of closeness is taken into account. In order to own an animal recognition is required. Where the continental European animal trials seem to reveal a notion of animal intention but in fact offer the reverse, the English law designates the animal as object but then enforces a closeness with the owner which the binary owner/owned seems to deny.

Domestic working animals were frequently put to pasture on common land, or, it would seem from the number of references to the practice found in depositions, were kept by a neighbour on behalf of their actual owner.[57] By placing the animals in common the danger of theft was increased, as was the potential for confusion over the ownership of a specific animal. In one case recorded in the East Sussex Record Office in 1642 an accusation is made that a sheep placed by its owner in the flock of a neighbour was slaugh-tered without the owner's permission.[58] There is no record of punishment, so it is likely that a recognition of genuine error was accepted in this case.[59] To attempt to counter such problems the animals were made recognisable, and in depositions dealing with sheep theft the references to ear clipping and wool marks are common. The animal is cut, painted, branded and labelled as property.[60]

Such markings, however, like the use of markings on hawks, are by no means conclusive. Cynthia Herrup records a case in which two neighbouring sheep owners used the same mark.[61] And in three cases of theft recorded in the East Sussex Quarter Sessions in 1644 and 1645 the alleged sheep thief is said to have altered them. In his confession to the theft of almost one hundred sheep over a two-year period John Boyes says of his partner, Edward Hide, that 'such sheep as should be brought into Hides Earmarke Hide had away with him'.[62] The change of earmark meant that Hide could pass the sheep off as his own. In a case from 1644 the deposition tells of the rediscovery of a sheep lost six months earlier:

> the said Langford being lame the said sheepe was brought in unto him and hee seeinge her knew her to be the sheepe of the said John Humphrey wch he kept for him and the marke was newly altered as likewise the lambe he believeth to be the same lambe that run away wth this sheepe.[63]

Similarly Abraham Bodle found two ewes and two lambs which he had lost in William Clarke's flock, and noted that 'the eare marke of his sheepe & lambs was newly cut out, and the woole marke all new plaistered over with tar'.[64] The changes in the marking reveal the impermanence of even the simplest operation of human reclamation.

But the other side of this is that in these cases where the marks no longer reveal the sheep the sheep were still identified. Where *scienter* emphasises the importance of recognising the behaviour of the animal, ownership requires another, but similar form of knowing. This recognition of the animal is repeated in other cases, and what is revealed is a sense in which the owners knew their flocks and herds as individuals. The importance of the recognition of the animal by its owner, what might be termed its individualisation, is enacted in the return of the lost sheep. This is reiterated in other cases where the change in the status of the animal – from sheep to mutton – reflects the attempt to remove the sheep's individuality. It is more difficult to recognise a skinned carcass than to recognise a sheep: it is an object, not a subject.[65]

In one case recorded in East Sussex in 1644 the skin of the sheep killed illegally by the accused's dog is removed and left in place of the sheep while the carcass is carried home and put into a sack under the bed. The search for the lost sheep becomes a search for meat, and less than a day after the disappearance only 'three quarters of mutton' is found.[66] In another deposition from 1648 Thomas Relfe tells how he found the skins of two sheep, one black and the other white, on Buxted Common, and two sets of footprints in the snow leading from the skins to the houses of Nicholas Sage and Nicholas Barden, one set entering each house. A search of the premises revealed confirmation of guilt.[67] The deposition reads:

> they searcht & found a hindquarter of mutton in the said Sages house neere a bedside, & a forequarter in the hedge hard by his howse, & 4 quarters more of the rest of the said 2 muttons, being half a black & a white sheepe, wch they discerned by some wooll left upon the meat they found in a sack, or bagge of the said Bardens in a litle Coppice close by his howse, & after comparing quarters, & chining of the sheepe together they found them to be pte of a black; and pte of a white sheepe divided equally betweene them in 4 quarters.[68]

In a case from 1644 the theft and dissection of another sheep is taken to further extremes:

> in the morning Greenes wife brought the sayd mutton into her house to her Dame Bryans wife & desired her to dress it for her & desired her to have it baked in a dripping pan in her oven who baked it in her oven & then Greene & his wife & George Pope did eat part of it & her dame did sett the rest aside untill the Constable came to search the house & then her dame put it in this Informers lap and bad her carry it into the Brewhouse & her dame came afterwards & take the mutton from her this Informer & layd part of it behinde fagots & put parte of it into the hogwash tubbe & all this was done the night and day before Mr Hatcher made search for mutton.[69]

From the delivery of the stolen mutton at midnight to the cooking of the meat the next morning, the journey from pasture to plate is a very short one. The constable and the owner know this and are not searching for the sheep but for mutton. In these cases the character and individuality of the animal is destroyed and the rediscovery of the (former) animal can only be achieved through the putting together of mutton jigsaws, the confession of theives – which are not always necessary in depositions in which the live animal is recovered – or eyewitness accounts. Recognition counts for nothing.

The distinction between the animal and its meat product is not only a linguistic one, it is a tangible, visual one, reflecting rather differently Thomas Adams' idea of Adam seeing the insides and outsides of animals.[70] In fact Adams himself seems to understand the difficulty of his image of absolute language in the face of meat-eating when, later in *Meditations Vpon Some Part of the Creed*, he notes that

> The killing of beasts, on what occasion soeuer, whether for food, for knowledge, or pleasure, belongs vnto the bondage of corruption, which sinne brought with it ... The beasts should not haue died for vs, if wee had not been dying in our selues. I am perswaded it was sinne that made vs butchers, and taught the master to eate the seruant.[71]

Adams' fears are repeated in another Reformed text: John Moore writes

> So in our meates (as in a looking glasse) we may learne our owne mortalitie: for let us put our hand into the dish, and what doe we take, but the foode of a dead thing, which is either the flesh of beasts, or of birds, or of fishes, with which foode we so long fill our bodies, untill they themselves be meate for wormes?

Through meat we 'taste [death] daily ... we feel it betwixt our teeth'.[72] There is no evidence to suggest that either Adams or Moore were vegetarians but the sense of the inherent sinfulness of meat-eating is antidote to George Herbert's Anglican anthropocentrism.[73] The Reformed sense of the lowness of humanity has implications for the place of the beast. If the beasts say 'eat me' what is to stop men from doing the same?

In his sermon, *Mercy to a Beast*, John Rawlinson wrote,

> It hath therefore ben a prundent *caution* of our Law givers, that *Butchers* (men acquainted with shedding the bloud of *beasts*)

should not be admitted for *Jurors* in cases of *life and death*: it being
a strong and violent presumption, that he that hath no pity upon
the life of *beasts*, will not have so much as he ought to have upon
the life of men.[74]

This problem emerges in an even more unexpected place in the late
1640s. In a letter dated *c*.1647–8 Robert Boyle contemplated the
issue of animal suffering and the ethical position of the scientist.
Boyle did not, despite some internal debate, regard his experiments
as being beyond morality.[75] He did, however, see the work of
another (less educated) anatomist as being dangerous to humanity:
'our English Lawes do wisely exclude Butchers From being
Jury men upon Life & Deth; as supposing their profession
but a Prenticeship of Cruelty'.[76] There is no statute to support
Rawlinson's and Boyle's claim, but it is reiterated and added to in a
very different text dating from 1655. The vegetarian Roger Crab
removes some of the blame from butchers and places it onto meat-
eaters themselves:

> I shall return to the reasonable part of the Law in this Nation,
> which excludes Butchers from being jury-men of life and death:
> Surely, if they are judged incapable of being a Jury, because they
> kill the creatures, they that buy it with their money to devour it,
> cannot be clear, for we alwaies count the Receiver more subtil
> and worse then the Thiefe; so that the buyer is worse then the
> Butcher.[77]

Through this legal tangle Crab raises further doubts about the rights
of humans to exploit and use animals. Eating meat becomes, in his
eyes, the equivalent to being accessory to a crime. Crab is, of course,
unusual in the nature of his stance – he was, according to the title of
the text, *The Wonder of this Age* – but the questioning of the rights of
dominion over animals is something that is implicitly produced by
the law itself, and this is, perhaps, why Crab uses this spurious yet
popular legal point to convey his argument. To possess an animal is
to be human, but to possess an animal is to give it a character; to
then go on and eat it raises some very uncomfortable questions.

The answer to these uncomfortable questions comes, in part, in the
removal of the working animal from the living environment, a move
which occurred during the sixteenth century. The animal was literally separated from its owner. This change in living arrangements

coincided with the opening up of the way for the other type of animal – the non-working, recreational creature – to be brought into the house.[78] Physical closeness to these animals could be maintained because the exploitation was very different: edibility was no longer an issue with a pet.[79]

IV

Having presented working animals as the equivalent of 'housholde-stuffe', William Lambard goes on to define the status of the non-working animal: 'to take dogges of any kind, apes, parats, singing birds, or such like (though they be in the house) is no *Felonie*: because these latter bee but for pleasure onely, and are not of any value'.[80] The difference in the use of the animal is represented in the law where, almost inevitably, the status of the creature depends on its 'profit' to man.[81] The law fails to recognise this recreational creature – does not give it a Latin tag – not because the recreational animal does not exist, but because it does not fit the legal notion of what an animal is for.[82] Instead of the Latin tag, however, the recreational animal does get a pet-name. It is introduced to the human community through a pseudo-Baconian rather than a legal process. John Taylor the Water Poet tells a comic tale about giving animals human names, a practice which was increasingly common in the seventeenth century.[83] An old woman complains about a man who has called his dog 'Cuckold'; 'out upon thee, thou misbeleeving knave ... where learnst thou that manners to call a dog by a christian bodies name?'[84]

The range of animals owned as pets seems to have been broad in this period: not only are the obvious animals such as dogs kept as pets, squirrels, apes and singing birds also enter the domestic arena.[85] More conventionally, however, Edward Topsell, using John Caius' words, refers to the ownership of lapdogs by 'dainty dames':

These puppies the smaller they be, the more pleasure they prouoke, as more meete play-fellowes for mincing mistresses to beare in their bosomes, to keep company withal in their chambers, to succour with sleep in bed, and nourish with meate at bourde, to lay in their lappes, and licke their lippes as they ride in their Waggons ...[86]

What emerges from the description is the potential for the sexual-isation of the relationship with the pet. The woman is presented in an unnaturally close relationship with the pet: these 'wanton Women' even 'admit [the dogs] to their beds, and bring up their young ones in their own bosomes'.[87] The relationship is based on women's 'corrupted concupiscences'; the animals are not presented as wild beasts requiring taming but as substitute humans.[88]

Such work concentrates on an elite possession of pets, and for women of the lower orders the problems existed in equal if differ-ent measures. There were links to witchcraft: 'the witch was likely to possess a familiar imp or devil, who would usually take the shape of an animal, usually a cat or a dog'.[89] The familiar was only ever found in English witchcraft cases. George Gifford relates the case of an old woman who confessed to owning:

> three spirits: one like a cat, which she called *Lightfoot*, another like a Toad, which she called *Lunch*, the third like a Weasill, which she called *Makeshift* … The Cat would kill kine, the Weasil would kill horses, the Toade would plague men in their bodies.[90]

In 1582 in St Osyth Ursula Kemp was hanged for witchcraft: her 'only crime was a malicious tongue, loose morals and a harmless friendship with two cats'.[91]

For women of the higher or the lower classes, then, having a pet was potentially dangerous. Attacks on female pet-keeping implied a number of things: that without a firm (male) hand the woman would make such a mental descent that she would disregard the differences between the species and show herself to be close to the beast; that she might, perhaps even more subversively, misrecog-nise the role of the man to such an extent that an animal is felt to be able to fill his place; and that, instead of reclaiming perfection, she would re-enact Eve's temptation by the Devil and make the animals not merely wild, but satanic.[92]

This is obviously deeply misogynistic, but the closeness of the human and the recreational animal is echoed in a very different arena. Within sporting manuals a similar sexualised relationship is proposed between the male hawker and his bird. Unsurprisingly, in these works the hawk is presented very clearly as female – the dom-inant partner in the training is definitely the man – but the manuals seem to take this a step further. Gervase Markham advises his gentlemen readers that:

All Hawkes generally are manned after one manner, that is to say, by watching and keeping them from sleep, by a continual carrying of them upon your fist, and by a most familiar stroaking and playing with them ... and by often gazing and looking of them in the face with a loving and gentle countenance, and so making them acquainted with the man.

After your Hawks are manned you shall bring them to the lure by easy degrees, as first making them jumpe to the fist, and after fall upon the lure, then come to the voice: and lastly, to know the voice and lure so perfectly, that either upon the sound of the one, or the sight of the other, she will presently come in and be most obedient, which may be easily performed, by giving her rewarde which she doth your pleasure, and making her fast when she disobeieth.[93]

The link between tameness and ownership makes it crucial that the hawk should know who to fly back to, and so through gazing into the face of the bird and ensuring it comes only to his voice the personalised relationship between the man and his hawk becomes central to the reclamation of the bird. A lack of authority would result in the loss of the property (it is important that the male hawker is able to *own* his recreational creature; the woman can never own her lapdog). The most explicit manifestation of this sexualisation of hawking is in Shakespeare's *The Taming of the Shrew*, where Petruccio says:

> My falcon now is sharp and passing empty,
> And till she stoop she must not be full-gorged,
> For then she never looks upon her lure.
> Another way I have to man my haggerd,
> To make her come and know her keeper's call –
> That is, to watch her as we watch these kites
> That bate and beat, and will not be obedient.[94]

The taming of the hawk and *The Taming of the Shrew* have much in common. Once again the woman is animalised.

These sexualised descriptions are not only found in manuals dealing with hawks which would, following the proscriptions on the ownership of hawks, be limited to members of the gentry or above; they can also be found in manuals dealing with cock-fighting.[95] Here, though, the animal is, of course, truly male; his warlike qualities are the stuff of panegyric. When George Wilson advises that wives should be made to 'love [their] husbands Cockes' the implications are clear.[96] The status of the fighting cock and the virility of the male owner are closely linked.

The 'ownership' of recreational animals is related to the changing place of the domestic animal in early modern society: the notion of character and individuality is carried from one sphere into another. As one type of animal leaves the human domain another takes its place. The actual separation from the exploited creature is an attempt to remove the difficulties which ownership – a recognition of individuality – enforces, and as such is an attempt, within Coke's definition, to assert difference, human-ness: the mere possession of the animal as property does not allow for this assertion. A literal wall is erected between the species. The pet is also an assertion of human-ness through the issue of naming, but because of the sexual-isation of the relationships which can be traced in contemporary writings, it is a dehumanisation of the owner as well. The closeness is written, created by the law – artificial reason – but in attempting to create humans the law instead creates bestialists.

I use the term bestialist as an antithesis of human here because that is how it was understood in the period. The bestialist lost his claim to human-ness when he performed an act which destroyed the boundaries between the species.[97] This is especially clear in the early modern belief that bestiality could produce monstrous off-spring. Just as the new science was destroying the distinction between the species through experimentation, so the continuing popularity of the old science meant that the possibility of cross-breeding was maintained. In committing the crime of buggery the human sank to the level of the beast and potentially produced a beast.[98]

In 1573 the French surgeon Ambroise Paré proposed that mon-sters could be produced through 'the confusion and mingling together of the seed'. Bestiality, he argued is a 'horrid' crime because from it 'Monsters have been generated and born, who have been partly men and partly beasts.'[99] The belief in cross-breeding can be traced in older sources, from the Greek myths to the early thirteenth-century work of William of Auvergne, who believed that bears produced semen compatible with human reproduction.[100] The myths remained powerful throughout the seventeenth century; in the 1670s the 'issue' of a bestial relationship between a man and a mare was 'nailed up in the Church-Porch of [Birdham near

Chichester], and exposed to publick view a long time, as a Monument of Divine Judgement.' The alleged bestialist, 'A Young Man of the Neighbourhood ... fled out of the Country.'[101] A similar desire to flee is recorded in an earlier deposition: John Sweedale recorded that William Clarke, who he had allegedly found 'committing Buggerie' with a mare, 'prayed him for gods sake to keep his Counsell and he would not stay two dayes in England.'[102]

The desire to leave the community found in these two cases reveals a voluntary exit from the humanising qualities of society: an echo of Orson's return to the forest. But if this type of exit was not achieved another one was found involuntarily. The bestialist had committed 'a sinne against God, Nature, and the Law', and was abandoned by all three.[103] In 1569 Edward Fenton asked if whoremongers receive such 'grievous paines, cruel punishments, irefull cursings sent by God ... what may then *Sodomites* ... hope for who joyne them selves ... with brute beasts[?]' He does not answer this question directly, but his punishment for whoremongers (those who are bad, but not as bad as bestialists) offers some sense of the irreconcilable nature of the worse criminal:

> they become blinde and madde, and after that they be once for-saken of God, and will not be reconciled by good and wholesome councel, but persever still in their wickednesse, provoking thereby Gods wrath and indignation against them.[104]

The bestialist by extension is lost to God, nature and the law: he is separated from Reformed, scientific and legal delineations of human-ness, and in the madness which Fenton predicts, fails within the humanist scheme as well. All notions of human-ness are cast aside in bestiality even as the law produces the need for closeness to animals. But madness also brings problems within the law. Where animal ownership creates the need for recognition which in turn emphasises a dangerous closeness, accusations of insanity also question the status of humans.

V

The official legal title of bestiality in the Middle Ages was *offensa cujus nominatio crimen est*, the offence it is a crime to name.[105] By refusing to name the crime which blurred the division between human and animal the law-makers refused it entry into the human community. Recognition was denied.[106] For Edward Coke, writing in the early seventeenth century, bestiality was still 'a detestable, and abominable sin, amongst Christians not to be named'.[107] The eliding of the terms 'sin' and 'crime' is important. The offence against the law is also an offence against God, and artificial memory gains divine sanction. A misdeed is 'not only harmful but also offensive .... crime remained closely linked to sin, deliberate evil, and moral weakness'.[108]

In this sense the fallen human within Reformed ideas was likely to sin.[109] Sinfulness was part of human nature, and the law was in place to act as a safety net to catch sinners on their inevitable descent. Calvin wrote: 'We must repudiate the opinion of those who hold that all sins proceed from preconceived pravity and malice. We know too well from experience how often we fall, even when our intention is good.'[110] In response to this natural depravity William Perkins proposed that there were two types of sin: 'either personall, or the sinne of mans nature'.[111] The division between these two is important and reveals the possibility of the distinction between willed sin (personal) and unwilled sin (natural). In turn, this allows the law to differentiate intended and unintended crime. The law emphasised *scienter* – knowing – in the ownership of an animal, and thereby implied intention: in the case of the dog-owner referred to earlier it was judged 'not lawfull to keep Dogs to bite and kill Swine' as if that was the only reason the man owned the dog. In the same way the judge had to distinguish between what was meant and not meant: he had to read the crime deeply. As the still influential thirteenth-century legal writer Henri de Bracton proposed in *De Legibus et Consuetudinibus Anglicae*, 'Theft is not committed without the thought of thieving.'[112]

However, there was another side to this. Coke wrote that 'no man may be punished for his thoughts. For ... thought is free.'[113] There must be an intention for a crime to have been committed, but there is also a sense in which intention is outside of the range of the law and is therefore not a crime. This paradox can be illustrated in two very different representations of bestiality in the early modern period. In an undated letter to the King's physician, William Drummond of Hawthornden (1585-1649) tells the story of:

> a poore miserable fellow accused of Bestiality; and he at his Arraignement confessed, that it was not out of any evil intention

he had done it, but onely to procreat a Monster, with which (having nothing to sustain his life) he might win his bread travelling about the country.[114]

In this instance the act is not denied, but the intention is confessed as innocent. The defence rests on the notion of pure volition: he didn't want to do it but he had to. The crime committed here is accidental, and therefore, the perpetrator seems to say, not as wicked as the same crime committed deliberately.

In another case of bestiality the opposite defence occurs. In response to William Stenton's deposition that 'hee saw John Swallow his fellow servant sett a gray Mare of his Masters with her hindde foot in a Gutter, and saw him jumpinge att ye Mare behind as a Horse doth', Swallow acknowledges that he did go 'behinde ye Mare, and thought to have Buggered her, but god gave him grace that hee did itt not.'[115] Swallow proposes that the intention to commit the crime, which he had, is of no importance, and that the act, which he denies performing, is the thing itself. His innocence rests not on his intention (which was inevitably sinful) but on his actions. The contradictory nature of these two cases reveals the difficulty in adjudging intent and also reveals a lack of clarity in the law.[116]

But the emphasis on the importance of human intention in committing a crime obviously differentiates a human from an animal who has no intention. As the deodand and *scienter* legislation discussed earlier show, the animal which kills is never guilty it merely represents a lack of responsibility on the part of the owner. A death caused by an animal is a casual death. Humans commit (perform) murder or manslaugher, animals (like inflated bladders) are the causes of accidents. Within early modern French legal ideas the animal in a case of bestiality is merely an 'instrument', an 'accessory': 'For it is necessary that all sin be voluntary and proceed from the act and shrewd resolution, which is wanting in beasts'.[117]

It would seem simple then to define the human as a creature who can act voluntarily and understand the criminality of his or her actions. Human-ness should be defined within the law by the capacity for intention. This would bring the notion of artificial reason to the fore. The human knows the law and lives within the law because it represents a standard of reasonableness, and a failure of this knowledge on the part of the human creates a loss of status. But Coke, of course, does not do this, he defines the human as having a certain physique. The possessor – that other distinctive position in the law, distinctive because an animal can never own or reclaim anything – has a human shape. The reasons for this intellectual leap lie in the problem of intention itself. Once again, just as the animal's position in Reformed ideas and humanism is the stable one, so in the law it is humans who constantly fail to exhibit human-ness.

In William Perkins' work the figure of the frantic was aligned with the animal.[118] Reformed ideas proposed that loss of mind meant loss of conscience, and loss of conscience meant loss of human status. A similar problem emerges in the law. *Non compos mentis*; the phrase can be translated as 'not in possession of one's mind'. It is the place where the definition of humanity as knowing and possessing breaks down. The law says that an idiot can inherit, but a human who is not in possession of their own mind (let alone other material possessions) is hardly to be defined as unproblematically human. Natural reason may be dismissed by the artificial reason of the law, but *scienter* implies that some form of reasonableness is still needed to exercise the right to property. The idiot crosses over the boundary between animal and human: he is not in possession of his mind, but is legally capable of possessing property; lacking in intention, but legally bound to fulfil deeds, feoffments or grants.[119] The 'wild beast test' which first emerges in Bracton is at the heart of the definition of the status of the idiot in the early modern English law and also represents the hybridity of this creature.[120] The idiot is neither human nor animal, but a human-animal in the truest possible sense.

This linking of the idiot with the animal reappears with the issue of intent. Echoing Bracton (cited earlier), Coke wrote, 'No felony or murder can be committed without felonious intent or purpose.'[121] This is reiterated by Dalton, who writes of idiots that 'they haue no knowledge of good and euill, nor can haue a felonious intent, nor a will or mind to do harme'.[122] A death caused by an idiot is like a death caused by an animal: it can only be understood as casual. But the parallel goes even further. Where the execution of animals in the continental European trials never served a direct admonitory function – no pig ever attended – so the execution of idiots was seen to have no function:

> the principall end of punishment is, That others by his example may feare to offend, *Ut poena ad paucos metus ad omnes perveniat* [in

order that the penalty may touch a few, but the fear thereof touch many]: But such punishment can be no example to Mad-men.[123]

Unable to reason, take responsibility, or read the execution, the idiot is an animal in the law. Like the animal in deodand legislation, the idiot is not executed: capital punishment is regarded as inappropriate.[124]

The legal definition of a person *non compos mentis*, again like the legal defintion of animals, has different categories:

1. Ideot or fool natural: 2. He who was of good and sound memory and by the visitation of God hath lost it: 3. *Lunaticus, qui gaudet lucidis intervallis* [a lunatic who is lucid at times], and sometimes is of sound memory, and sometimes *Non compos mentis*: 4. By his own act, as a drunkard ...[125]

The drunkard is drunk 'By his own act' (drinking is an act of will) and is therefore considered capable before the law: in fact Perkins' explanation for the inactivity of conscience does 'not extenuate [the drunkard's] act, or offence, nor turn to his avail, but it is a great offence in it self, and therefore doth aggravate his offence'.[126] The 'Ideot *à nativitate*', a 'natural fool', is very different from both the newly made madman and the intermittent idiot. And while those who fit into both of the latter categories of madmen may possess property, the property of the 'Ideot *à nativitate*' has a different status: it is given in custody to the king who 'is bounden of right by the Laws to defend his Subjects, and their goods and chattels, lands and tenements'. But, significantly, 'the King hath not the freehold or fee, but the freehold is in the ideot'.[127] The idiot possesses the property for life although the control of it is in the hands of the monarch. The idiot is different in this from a ward, an under-age heir, in that it is not only the ward's inheritance which is received by the guardian, but his body as well.[128] The idiot is in possession of his own body: he retains, if you like, human shape.

The body, the very thing which the new science had to dismiss as a signifier, returns because there is no other logical definition of the human available in the law. Responsibility in ownership, like naming in science, should signal dominion, but in the law responsibility can only be achieved through *sciens*, knowing the animal. This creates the need for a closeness, and this closeness brings the owner, the human, close to being a bestialist, an anti-human. In the alternative definition of human-ness, intention signifies a criminal act – Coke wrote 'if it be voluntary, the law implyeth malice' – and crime is something which is committed only by humans (an irony if ever there was one).[129] Animals, objects in the eyes of the law, and idiots commit acts without intention and therefore cannot be criminals. But idiots, like truly human humans, can own property. The lines which separate humans from animals have become blurred, and the body, that most concrete and problematic of signifiers, is used, with no religious connotations, to define the species. Looking like a human is what defines the right to possession, and possession is dominion in a fallen world. What is revealed as animal in the new science defines the human in the law.

But, of course, ownership and status, like the work of Edward Coke itself, return to centre stage in the 1640s. Mine and thine – the linguistic designations of property rights – emerge as crucial ideas in political debates, but so does the status of the human itself. In the work of the Leveller Richard Overton, the focus of the next chapter, the two are brought together. What emerges in Overton's work is an extension of Coke's. Overton articulates the fragility of the body as a signifier of humanity and looks to political change for an answer. The animal is again central, but a form of equality between the species – impossible in Coke's scheme – emerges.

6

The Bestialisation of Humanity and the Salvation of the Beast: The Politics of the Animal Soul

In early modern culture the notion of an *a priori* humanity, a separate and distinct category, is continually under threat from the beasts which seem to support it. Qualities of human-ness rely on the presence of the animal, but where there is an animal these qualities which seem to define what is human about the human are revealed to be beastly. This loss of status is evidenced in an anxious anthropocentrism which results in the increasing exploitation of animals. In order to declare superiority humans stage baitings, deny animals access to God, objectify them in fables, place them on the anatomist's table, mark and brand them as possessions. But in each exercise of dominion the antithetical position emerges: humans become the animals they attempt to dominate.

Dominion and the anthropocentrism which it voices can never confirm or establish the status of humans and another way of relating to the animal is needed. A proposal for this new relation can be found in the work of Richard Overton.[1] Where the animal in previous formulations operates as a spectacular, religious, educational, scientific and legal way to delineate the human, in Overton's work it is political. The animality of humanity which this book has traced is understood as the result of political misrule and Overton's search for a new human is inseparable from his search for a new form of government. But he goes further than this. To regard the animal as a means to confirm human status is to support the anthropocentric vision which declares that the animal is only important for what it can tell humans about themselves. Overton refuses this logic and he recognises the animal as an animal, important in and of itself.

143

Whether in government or in the natural order, dominion destroys human status, and Overton's assertion of a new humanness based on self-propriety rather than oppression is paralleled in the assertion of the equality of all animals. The links made between the political and the natural order are to be found in Overton's early work, *Mans Mortallitie*, and the relocation of this text in the Leveller canon is the starting-point of this final chapter.[2] What follows argues that Overton's declaration of humans' and animals' shared fate in the afterlife is related to the call for political change in his better-known Leveller writings. The new human which Overton is seeking is a product of a new relation to power and to the animal.

I

The attribution of texts written during the English Civil War is notoriously difficult. Authors used pseudonyms, they printed work without attribution, and were parodied. Richard Overton is no different from many other writers of the period. While some of his works are easy to attribute in that they are autobiographical or signed, others have caused more problems. *The Arraignment of Mr Persecution*, for example, is now read as an Overton work but was first printed under the name Martin Mar-Priest – Overton's tribute to the pseudonymous Martin Mar-prelate of 1588–9. *Mans Mortallitie*, first published in January 1644, appeared under the authorship of R.O. and the title page stated its place of publication as Amsterdam.[3] This led to the suggestion of the authorship of Robert Overton, one-time Governor of Hull, but this attribution of the text is flawed. On his edition of *Mans Mortallitie* Thomason crossed out Amsterdam and wrote instead 'London', and, as Perez Zagorin notes, the naming of John Canne as printer is also false, a deceit used by 'the author of so heterodox a composition … to evade identification'.[4]

The text can be attributed more certainly to Richard Overton, not only because *Mans Mortallitie* is referred to in *The Arraignment of Mr Persecution*, but also because some of the ideas which it proposes can be traced directly in another of his works.[5] But on top of these links between texts, the attribution can be confirmed through a detailed analysis of its contents, and a recognition of the links between the views which this text announces and the ideas which

Overton proposes in his later work. *Mans Mortallitie* is concerned with the status and nature of the human soul, but its implications are wholly political. What follows here is a brief outline of the text. Some of the ideas which are summarised, most notably mortalism and the natural link of the body and the soul, will be discussed in more detail later in the chapter.

Mans Mortallitie first appeared in an uncorrected edition dated by Thomason as 19 January 1643/4. A second, corrected edition appeared later in 1644. In 1655 an enlarged version entitled *Man Wholly Mortal* appeared which was reprinted in 1675.[6] In its original form the text is fifty-six pages long, and divided into seven chapters. Beginning at the beginning, Chapter I is entitled 'Of Mans Creation' and examines the implications of Adam's birth and the Fall. God's warning to Adam in Genesis 2: 17 ('But of the tree of the knowledge of good and evil, thou shalt not eat of it: for in the day that thou eatest thereof thou shalt surely die.') is used to propose the mortalist heresy. If the wages of sin are death, then, Overton argues, 'the Souls possibility of sinning being producted into Actuall sinne, the Soul must have its *wages*, Actuall mortality.'[7] Overton proposes that the human soul dies with the body and that it rises to heaven only on Judgement Day.

In Chapter II, 'Scriptures to prove this Mortality', Overton uses biblical citations to defend his belief, and writes

> if Death be not dessolution of life, or its depravation, how can it be said to suffer death? Not by a bodily separation, for that is but as the laying downe of a burthen, where with it was clogged and tyred, whereby it is made only more lively ten thousand times; (as my Opposites confesse;) and so, can no more be said to be dead, then a *Porter* when he is disburthened of his load.[8]

The soul and the body must both be mortal in order for death, the great punishment, to be truly experienced. The period of the death of the soul – the time from the death of the individual to Judgement Day – is not measurable as time because in death the individual 'absolutly *IS NOT*'.[9] Overton writes later in the text that 'the twinkling of an eye to the living, is more time; then a thousand, yea ten thousand yeares is to the dead: For *Being* only commensurates with Time, or length of dayes, *not to Be* cannot possibly be capable there of.'[10] He adds, 'There is nothing so represents death, or *non-being* as *sleepe*.'[11]

Chapters III and IIII of *Mans Mortallitie* turn away from the Bible and offer evidence of the soul's mortality in the work of 'Naturall Reason'. Aristotle, Nemesius and the late sixteenth-century French surgeon Ambroise Paré are all invoked to support Overton's belief that the soul is 'all the internall and externall Faculties of man joyntly considered.'[12] Only after physical evidence has been found does Overton then return to the Bible in Chapter V, the title of which sums up its purpose: 'Objections extorted from Scripture Answered'.[13] This chapter, the longest in the work at twenty pages, is followed in Chapter VI by a return to the arguments of natural reason, and a discussion 'Of procreation, how from thence this Mortallitie is proved'. Overton disputes the presence of the soul at the first moment of conception worrying about the status of the miscarried embryo. Rather than assert the immortality of 'that *Effluction* or vnshapen deformed peece of *congealed blood*' Overton proposes that 'as the whole Tree is potentially in the seed, and actually in time springeth from it ... so in the seed of mankind, is whole man potentially, and wholy actuall in time'.[14] Utilising the language of both Reformed and humanist educationists, Overton sees the embryo as a potential rather than an actual human, and by extension, the soul developing with the development of the body. Finally, and very briefly, Overton offers 'Testimonies of Scripture to prove that whole man is generated, and propagated by Nature'.[15]

Until now *Mans Mortallitie* has been seen as a marginal Leveller text, if it is considered as a Leveller text at all.[16] Harold Fisch proposes that it was 'written prior to [Overton's] major involvement in political causes .... It lacks any political aim or context, and such political references as there are are incidental to its major interest in the fate of the human soul.'[17] But Christopher Hill's argument that 'the division we draw today between "religion" and "politics" does not apply in the seventeenth century' can be used as a starting-point for a reassessment of the work.[18] The religious context is central to an understanding of mid-seventeenth-century political ideas: H. N. Brailsford proposes that 'the Levellers can hardly be understood without reference to the Anabaptist tradition that lay behind them. The party was translating into a political programme the ethics of a religious movement.'[19] It is from the baptist tradition that *Mans Mortallitie*'s role as a political as well as theological document can begin to be assessed.

Richard Overton was a General Baptist. That is, he was a member of a congregation which believed that Christ died for all, that there was a 'general redemption'. Overton's ideas differ from the predestinarian ideas of Reformed theology which emphasise the possibility of reprobation as well as of salvation to the extent that they have been termed 'urban Pelagianism': there was 'a heavy emphasis upon individual responsibility in the matter of salvation'.[20] Overton illustrates this principle in *Mans Mortallitie*:

> none can be condemned into Hell, but such as are actually guilty of refusing of *Christ* because immortality or the *Resurrection* cannot be by Propagation or succession, as mortallity from *Adam* to his Issue, and so the Child, though temporally, yet shall it not eternally be puneshed for his Father's sinne, but his Condemnation shall be of himself.[21]

In contrast to Calvin's assertion of the innate sinfulness of all children Overton proposes that sin is something which humans perform. All individuals, the General Baptist argument goes, receive grace, but the individual must gain salvation through their own actions. It is in human choice that salvation is achieved: the human can refuse Christ. Such a choice is inconceivable within Reformed ideas.

His Baptist principles also distinguish Overton's beliefs from Reformed debates about the status of the infant in another way, and emphasise, following John Murton, one of the early leaders of the Baptist Church in England, that 'Infants cannot heare, nor beleeue. For they know not God, and then how can they beleeue, seeing *Faith* comes by hearing and is grounded vpon *knowledge*.'[22] The Baptist refusal of infant baptism brings into play ideas of infant innocence and of infant choice.[23] It also utilises ideas which can be traced in humanism: hearing, understanding is what makes the Christian a true member of the faith. The Reformed notion of the catechism as the method of educating the innate faith of the individual (the idea of double justification) is replaced by the belief that true adult understanding and *acceptance* of doctrine is necessary for full entry into the church. Countering Murton's ideas, John Robinson, the separatist who was to be the priest on the *Mayflower*, stated that 'it is absurdly said, that a man is made a *Christian by his Baptism* ... He that is not a Christian before he be baptized, becomes not one thereby.'[24] We are back to the significance of the conversion of Nathaniel the Jew *before* his baptism (discussed in Chapter 2). For Reformed thinkers salvation is a gift from God; for General Baptists, however, grace is given to all, but salvation can be achieved and is based upon human choice. These two points are united in Murton's defence of the Baptist rejection of infant baptism:

> none shall appeare before Christ to receiue iudgement, but those that haue done workes, either good or euill, and that in the flesh. Infants dying, haue done neither good, nor euill in the flesh. Therefore infants shall receiue no iudgement.[25]

Unlike the Reformed emphasis on the predetermined nature of human salvation, the General Baptists believed in the duty of the individual and their role in their own salvation.[26]

Another accusation made against General Baptists was the belief in mortalism.[27] Overton was, according to Thomas Edwards, present as a moderator at a 'Disputation' about 'the Immortality of the Soule by some Anabaptists, as *Lam*, *Battee*, and others' in January 1646.[28] The heresy of mortalism was especially dangerous because it was the enemy within: 'developed and supported by Christians within the framework of Christian thought and feeling'.[29] In 1577 Heinrich Bullinger attacked the belief in no uncertain terms:

> If as yet there be any light-headed men to whom the immortality of the soul seemeth doubtful, or which utterly deny the same, these truly are unworthy to have the name of men; for they are plagues of the commonwealth, and very beasts, worthy to be hissed and driven out of the company of men. For he lacketh a bridle to restrain him, and hath cast away all honesty and shame, and is prepared in all points to commit any mischief, whosoever believeth that the soul of man is mortal.[30]

Bullinger, Zwingli's successor in Zurich, spells out the Reformed position. The soul is the seat of both reason and faith, to deny its immortality is to deny the very thing which differentiates human from animal. The conscience is placed in the human mind, which itself is inseparable from human immortality. To lose one – immortality – is to uproot the other – reason. To deny the immortality of the soul is to reveal a lack of bridling, education. A mortal human is not a human at all; he is, as Bullinger notes, an animal.

Overton counters this belief head-on in *Mans Mortallitie*. He writes, 'But Reason, vnderstanding, &c. may be absent from the

Body their Subject, and yet the Body living: as in made [*sic*] men, and persons in the *Falling-sicknesse*; and none will deny they are men at that same time.'[31] For Overton, as for Coke in 'Beverleys Case', reason cannot be used distinguish human from beast because reason can be absent in those who are recognised as human.[32] A declaration of mortality does not mean, therefore, a declaration of animality because the soul can never be the site of human ness. Overton continues:

> If [the soul] be the *Rational Facultie*, then all men are borne without *Soules*, and some die before they had *Soules*, as *Infants*; and some after their *Soules* is gon, as *Mad-men* that live and perish in their *madnesse*; and some would be borne, live, and die without *Soules*, as *Fooles*; and some would have *Soules* but by *fits* and *jumpes*, as *Drunkards*, persons with the *Falling-sicknesse*, &c. nay all of us spend a great part of our dayes without our *Soules*, for while we are in sound sleepe our *Rationality* ceaseth *pro tempore*: Thus this immortall spirit goes and comes as occasion serves.[33]

The dualism which was established by Reformed thinkers is destroyed by Overton. He proposes that the definition of humanity based on a division of body and mind, the corporeal and the incorporeal is an impossible one and cites, of course, the drunkard to support his claim. The physicality of the soul is essential: Overton writes 'wee see, that the *Eye* is no *Eye* without the *Sight*, and *Sight* no *Sight* without the *Eye*'. Vision, the immaterial act, is inseparable from the organ, the material instrument of vision, the eye, and vice versa.[34] Using the ideas of natural reason, he asserts that the soul is in the body, and the body is a part of the soul, and that this inseparability is traced in death as well as life.

But Overton's heresy does not end here. Having contradicted Reformed dualism, and undermined the humanist interpretation of human-ness which ignores the body and cites reason as the essential quality, Overton asserts that the separation of the human from the animal in the afterlife is also impossible. If inherently sinful humans are saved in spite of their sinfulness then how can the animal, that which has not sinned, suffer the fate of everlasting death? If a sinless infant is not to be judged, then neither is a sinless beast. The logic of Overton's position reveals not only the absolute heterodoxy of *Mans Mortallitie*, it also begins to uncover the ways in which we can link the animal with the human in his political ideas.

II

Writing in 1622 in an extreme rendition of the orthodoxy of Perkins, Rawlinson and Draxe (see Chapter 2), Godfrey Goodman proposed that animals have an innate and natural love of God, and that they therefore operate as an 'aide' to humanity. They 'beare witnesse [to the] ... admirable attributes in God' in order that humans may worship more fully: 'as there is a religion above man, the religion of Angels, so there may be a religion beneath man, the religion of dumbe Creatures'.[35] Unlike George Herbert's belief that animals were *unable* to worship God, it is, for Goodman, the original perfection which the creatures still exhibit – they 'continue innocent in themselves' – which allows them to reverence the Almighty as 'the praises of God require no more in effect, then the power and ability wherewith God hath first inabled the Creature'.[36] The notion of inherent faith found in Reformed ideas is mirrored in the innate perfection of animals, and means that, for Goodman, they can remind humans of their duty to God. By 'speak[ing] their Maker' the animals make an invisible God visible.[37] But Goodman goes even further than this:

> as they were ordained for his naturall use, for his food, clothing, labour; so it would seeme, they were appointed for his spirituall use, to serve him in the nature of Chaplaines, that they should honour and praise God, while their Master, sinfull and wretched man, dishonours him.[38]

In proposing animals as 'Chaplaines' to humanity, giving them roles as intermediaries between human and divine, Goodman is, in effect, replacing Perkins' idea of the conscience with the animal. This chimes in with Goodman's known recusancy: he continued to practise from the Roman breviary and in his will of 1656 he acknowledged 'Rome to be the mother church'.[39]

It would be simple to see a clear distinction between Reformed and recusant interpretations of the animal, but such a division is not in evidence. Rawlinson, for example, uses the Thomist interpretation of the beast in his work, while the worship of the dog-saint Guinefort continued in Catholic peasant culture until the late nineteenth century.[40] In Goodman's work the seemingly heterodox interpretation of the role of animals within the Christian faith is curtailed in a way which brings his text into line with more orthodox representations:

> what can be more glorious to God, then that his praise should be set forth by all his Creatures? what greater charity, then to comprehend them, not within the walls of our Christian Church (though once they were in the Arke) yet within the compasse and circuit of religion?[41]

The animal is a symbol of true worship, but it is not capable of true worship itself (Herbert's position). It is within the compass of religion but is excluded from the church. Goodman, like Draxe, Perkins and Rawlinson before him, sees the animal as being solely for human use. What he does is give this a recusant interpretation: the animal does not only, like a good Reformed preacher, remind humanity of the Fall; instead, like a good Catholic priest, it also reminds humanity of the generosity of God and the possibility of salvation.

It is easy to see how far Richard Overton departed from orthodoxy in *Mans Mortallitie*. Put simply, his logic is that if animals did become wild because of the Fall of man then they remain sinless, and the resurrection would have to include them for it to be truly successful. There could (theo)logically be no difference in outcome for animals' and humans' souls.

> the Beasts were not given man to eate in the *Innocency*, but to all fleshe wherein was *the breath of life, was given the greene herbe for meat*: Therefore, the death of the Beasts &c. was part of the *Curse*, and so to be done away by Christ. ... If the other Creatures do not rise agine, then Christ shall not *conquer death*, but when it is said, *O Death where is thy sting, O grave where is thy victory*? it will be answered in Beasts, because they are still captivated vnder its bondage ... Therefore Death shall not retaine them: but they must be delivered out of its Jawes.[42]

The wildness of animals which is so much a part of the legal debates about ownership is not a result of animal sin but of human. Calvin wrote,

> For GOD not being contented with the punishement of men, proceeded in taking vengeaunce vppon beastes, vpon fowles, and

vpon all kind of liuing creatures that are vppon the earth. Wherein he seemeth to punishe beyond all measure. ... those things which were created for mans sake, & which liued not to his vse, perished with him: and no meruell. The Asses and the Oxen with other beastes had in nothing offended: but because they were subiecte vnto man, he falling, they were drawen also into the same destruction.[43]

This was reiterated in a more concise way in 1633 by Henry Vesey: 'The Creatures were not made for themselues, but for the seruice and vse of man: and therefore the euill that is now come vpon them, is not their punishment properly, but a part of ours.'[44] The animal is the truly innocent victim of the actions of humans.

In many ways the opening premise of Overton's notion of the shared fate of humans and animals in the afterlife is based, as all General Baptist principles are, on orthodox ideas such as those held by Thomas Draxe.[45] His sense of the inseparability of humans and animals, again like other Baptist principles, then departs from the orthodox line in a significant way. Where Overton seems to return to the orthodox stance on animals he always takes it a stage further. For example, Overton makes an apparently customary distinction between the human and the animal in the here and now by stressing degrees of difference in what he terms the faculties: these range from the five senses to the more complex abilities of 'Reason, Consideration, Science, &c.'[46] He then, however, takes this differentiation to its logical extreme:

if it be not vnnaturall that *Seeing, Hearing,* &c. should be producted by an *Elementary operation,* as none deny in the propagation of *Beasts:* why is not the *Rationall Facultie* in *Man* as naturall in *Man,* and may as well be producted *elementarily* by *Man* as the other by *Beasts,* and be as actually mortall?[47]

Human reason, like animal sight and hearing, is produced elementarily: but humans have reason and animals have sight and hearing, and the difference between the two seems to be important. But Overton writes:

if the said *Faculties* in an inferiour Degree be elementary, so must they in a superiour: But ... *Brutes,* whom none deny to be wholly

Mortall, and all their *Faculties* elementary, have our most noble partes and *Faculties* scattered amongst them though in an inferiour Degree ...[48]

There appear to be differences between the species, but what is revealed is that these differences are not innate but are created, are potential rather than actual; 'But all mans Faculties, yea those of *Reason, Consideration, Science*, &c. all that distinguish Man from a Beast, are augmented by Learning, Education, &c. lessened by negligence idlenesse, &c and quite nullified by madnesse'.[49] He later writes, 'Therefore *Faculties* increase with their *Subjects*, and if increase, they must decrease.'[50]

The faculties which differentiate human from animal for Overton echo three areas which this book has looked at, reason (humanism), consideration (theology) and science, but all are presented as dangerously unnatural: they have to be augmented by an external addition like education. Implicitly, if not quite explicitly, the human is nothing more than a reasoning, considering and scientific animal. This can be taken even further: Overton is proposing that the qualities of human-ness become the substitutes for the human. There is an incomplete being which can be termed human, and this being is completed by learning which can, for Overton, as for humanist writers, be lost. The *a priori* human has disappeared from view.[51]

This difficulty of separating human from animal is further underlined by Overton in *Mans Mortallitie* when he proposes the 'Creatureship' of humanity: 'Humanity though glorifyed is but a Creature'.[52] The animality of humans seems to be explicit here. William Harvey's anatomical work placed man among the '*Creatures*' on the experimenter's table. William Perkins, while using the term to signify animals specifically in some works,[53] also uses it to describe humans. In his catechism *The Foundation of the Christian Religion Gathered into Six Principles* Perkins writes, 'If damnation bee the reward of sinne: then is a man of all creatures most miserable.'[54] This usage is a convention within Reformed works which emphasise the degraded status of humans; in fact, the double meaning of creature is what makes it useful. When Thomas Dekker wrote of 'a company of creatures that had the shapes of men' in the Bear Garden the term was deliberately and meaningfully degrading.[55] It calls upon the reader to recognise the ease of the movement between the two categories.

Both usages of 'creature' – designating the animal alone, and a degraded humanity – can be traced in the work of the Digger Gerrard Winstanley. In *Light Shining in Buckinghamshire*, he writes:

> and the creature Man was priviledged with being Lord over the other inferiour creatures, but not over his own kinde; for all men being alike priviledged by birth, so all men were to enjoy the creatures alike without propriety, one more then the other....[56]

What is significant here is that Winstanley maintains a sense of human dominion over the natural world: 'everyone was made to be a lord over the Creation of the earth, cattle, fish, fowl, grass, trees, not any one to be a bond-slave and beggar under the Creation of his own kind. *Gen.i.28.*'[57] For Overton, on the other hand, if his political works are read through the ideas contained in *Mans Mortallitie*, such unquestioned dominion becomes impossible.[58]

This element of Overton's ideas was recognised in contemporary criticisms of his belief in the equality of humans and animals in the afterlife. In 1645, in *The Prerogative of Man*, the author, echoing Bullinger, calls Overton 'a sorry Animal' and complains of his and his supporters' 'dishonour and debasing of their own kind, not elevating Beasts to the degree of reason ... but contrariwise reproachfully depressing man even as low as bruit beasts, and ascribing to them both a mortality alike.'[59] The significance of this attack on Overton lies in the equal emphasis placed by the author on the mortalist heresy and on the place of animals. The beast is not regarded as a side issue but is seen as central. The author of *The Prerogative of Man* sees Overton as a heretic and a betrayer of human nature: 'was it not enough that all inferiour creatures doe rebell against us, but we must basely and treacherously conspire against ourselves?'[60] But in making this criticism of Overton the author has not taken account of the reason for the animals' rebellion – humans – and this misunderstanding of the role of innocence is vital to his criticism of Overton. The author of *The Prerogative of Man* proposes that the animals become unnatural rebels against natural man, but this is the exact opposite to Overton's argument, which is that it is humans who have become unnatural and have created animal wildness.

Similarly, a year later, in his collection of 'Errours, Heresies, Blasphemies and pernicious Practices of the Sectaries of this time', Thomas Edwards notes the absurdity of the ideas that 'God loves the creatures that creep upon the ground as well as the best Saints; and theres no difference between the flesh of a Man, and the flesh of a Toad' (an idea which recalls the problems faced in vivisection); that 'The soul of man is mortall as the soul of a beast, and dies with the body'; and he also records for criticism the contemporary heretical belief that:

> There shall be in the last day a resurrection from the dead of all the bruit creatures, all beasts and birds that lived upon the earth, every individual of every kinde of them that died shall rise again, as well as of men, and all these creatures shall live for ever upon the earth.[61]

Overton, however, inverts Edwards' criticism that he is destroying the difference between 'the flesh of a Man, and the flesh of a Toad' in the Mar-Priest tract *The Arraignment of Mr Persecution*. This was written before the appearance of *Gangraena* and calls on Edwards – whose views were obviously already well known to Overton – to 'come forth in print against [*Mans Mortallitie*]'.[62] In this text Overton condemns Edwards (Mr Persecution) through the character Creation, who says,

> from the *Evidence* of the *Witnesses* against PERSECUTION, I clearly perceive, that he is of so divelish and unnaturall a disposition, as is not computible with the workes of the Creation, all creaturs of a kind associate, feed, and converse together, there is a publicke freedome of all kinds amongst themselves, the *Oxe*, the *Asse*, the *Sheep*, and all sorts of *Cattell*; the *Dove*, the *Sparrow*, all kinds of *Birds* have a State harmony, a publicke Toleration, generall Concord and unity among themselves in their severall kinds: but this *Malefactour* as from the *Witnesses* evident, unnaturallizes *mankind* above all kinds of Creatures, that where he rules, no peace publike, or private; no freedome, rights or liberty either civil or spirituall; no society, cohabitation, or concord Nationall or Domesticke can possibly be amongst men, but *envying, hatred, emulation, banishment* &c. Wherefore, from the consideration hereof, and of what the Witnesses have given in, to me he appeareth *guilty* ...[63]

It is Edwards and not Overton who, in his divisive intolerance, is the destroyer of humanity and human society.

The significance of the human in creating the status of animals which Overton presents through the narrative of the Fall, or through the bigotry of people like Edwards, is then translated into the political world. Worldly political order, he argues, destroys humanity, makes the achievement of difference between human and beast impossible. In making this move Overton is not betraying humanity, as the author of *The Prerogtive of Man* argued; it has, in his account, already betrayed itself. Rather, he is recognising and attempting to rectify an intolerant and intolerable situation.

For there to be true and fair order there must be a recognition of the false and unfair place of animals. The declaration that beasts partake in the afterlife is a political gesture: mortalist beliefs and Leveller principles are united in the animal. But the natural order, the existence of human dominion, is not merely a figure of the unnatural political order of England, it is absolutely inseparable from it. In politics the governors enact dominion and reduce the governed to the status of beasts. It is Overton's argument that to change one we must change the other: governmental order cannot be separated from natural order.

III

In the preface 'To the Reader' in the fifth part of his *Reports*, Coke presents the vital importance of the common law to all Englishmen:

> The au[n]tient & excellent lawes of England are the birth-right and the most au[n]tient and best inheritance that the subiects of this realm haue, for by them he inioyeth not onely his inheritance and goods in peace and quietnes, but his life and his most deare Countrey in safety. And ... I feare that many of my deare Countreymen, (and most of them of great capacity, and, excellent parts) for the want of vnderstanding of their own euidence, doe want the true knowledge of their auntient birth-right in some points of greatest importance ...[64]

Through the inaccessibility of the law humans – even those with intelligence – lack understanding and cannot possess, be human, with any sense of security. The systematisation of the law was Coke's attempt to rectify the situation. The law, in the middle of the seventeenth century, is political. But more than this, Coke's representation of what the loss of this birthright means is an obvious precursor to Overton's ideas.

In *An Arrow Against All Tyrants and Tyrany* Overton emphasises the essential qualities of humanity which originate at birth:

> To every individuall in nature is given an individuall property by nature, not to be invaded or usurped by any: for every one as he is himselfe, so he hath a selfe propriety, else could he not be himselfe, and on this no second may presume to deprive any of, without manifest violation and affront to the very principles of nature, and of the Rules of equity and justice between man and man; mine and thine cannot be, except this be: No man hath power over my rights and liberties, and I over no mans; I may be but an Individuall, enjoy *my selfe* and my selfe propriety, and may write my selfe no more then my selfe, or presume any further; if I doe, I am an encroacher & an invader upon an other mans Right, to which I have no *Right*. For by naturall birth, all men are equally and alike borne to like propriety, liberty and freedome, and as we are delivered of God by the hand of nature into this world, every one with a naturall, innate freedome and propriety (as it were writ in the table of every mans heart, never to be obliterated) even so are we to live, every one equally and alike to enjoy his Birthright and priviledge; even all whereof God by nature hath made him free.[65]

The political situation is a logical and painful extension of the arbitrary status of humanity in theology. Without the salvation of animals human salvation is meaningless; without the overthrow of the current political order all political order is useless: 'for tyrany, oppression and cruelty whatsoever, and in whomsoever, is in it selfe unnaturall, illegall, yea absolutly anti magisteriall, for it is even destructive to all humaine civill society, and therefore resistable'.[66]

Within the context of these ideas it is both wholly appropriate and inevitable that the human cannot be distinguished from the beast. The birthright, the self-propriety which is innately human, has been removed by the politicians. Political change is needed to restore not only right rule, but the status of humanity itself. However, the notion of equality which lies at the heart of Overton's political ideas seems to have its limitations: as C. B. Macpherson has pointed out, 'the Levellers consistently excluded from their franchise proposals two substantial categories of men, namely servants or wage-earners, and those in receipt of alms or beggars'.[67] The liberty which was proposed by the Levellers was an exclusive liberty: this was no true levelling, but only partial, and if freedom from arbitrary government is the freeman's birthright then it would appear that the Levellers, Overton included, are not actually seeking a total restitution of human rights, but only a partial one. This would make my claims for Overton's interest in the salvation and freedom of *all* somewhat difficult to make. But the exclusions in the franchise relate to Overton's understanding of government and his assessment of the postlapsarian nature of rule speaks of exclusions but also the possibility of freedom for all. Again, it is via the animal that such an argument can be made.

In *A Defiance Against All Arbitrary Usurpations* Overton presents a picture of humanity which offers a link between his theology and politics and, I suggest, reveals another way of understanding his ideas about the franchise. The passage is also clearly an echo of Coke's defence of the common law:

> Yea, such hath been the misterious mischievous subtilty from generation to generation of those cunning Usurpers, whereby they have driven on their wicked designes of tyranny and Arbitrary domination, under the fair, specious, and deceitful pretences of Liberty and Freedom, that the poore deceived people are even (in a manner) bestiallized in their understandings, become so stupid, and grosly ignorant of themselves, and of their own naturall immunities, and strength too, wherewith God by nature hath inrich'd them, that they are even degenerated from being men, and (as it were) unman'd, not able to define themselves by birth or nature, more then what they have by wealth, stature or shape, and as bruits they'll live and die for want of knowledge, being void of the use of Reason for want of capacitie to discern, whereof, and how far God by nature hath made them free ...[68]

This passage, written three years after *Mans Mortallitie*, reveals the impossibility of being human within the current political climate. It is because of 'tyranny and Arbitrary domination' that human-ness has been lost: men have been unmanned. But for Overton, as for Prynne before him, it is not in effeminacy that humanity is truly lost, but in animality.

What is also obvious in the passage from *A Defiance* is Overton's indebtedness to some of the debates which this book has already examined. The bestialisation of the understanding which he notes reiterates the humanist emphasis on judgement. A lack of discernment – as Sidney and Jonson proposed – is at the heart of animalisation. What is left is the ability to judge like a malt-horse, a thoroughly bestialised understanding. Such a sense of judgement is revealed in the anonymous parliamentarian play-pamphlet *A Dialogue Betwixt A Horse of Warre, and a Mill-Horse*, in which the political situation is discussed by two very different animals. The Mill-Horse says 'Thou pampered Jade that livst by plundered oates, / My skin's as good as thine and worth ten groates.'[69] There is theft, but there is also self-worth. The Mill-Horse speaks the Leveller position.

In the passage from *A Defiance* the 'specious, and deceitful practices of liberty and freedom' which Overton emphasises can be related, perhaps, to Sidney's interpretation of the Fall where a similar sense of a lost community can be traced. Sidney's beast fable shows how stewardship became dominion and shared, communal authority was replaced by deceitful tyrannical domination: 'Not in his sayings "I" but "we"; / As if he meant his lordship common be.'[70] In political terms the usurpation of power by the few in this fable is represented by the giving up of the voice, the ability which in the 1637 text of *Valentine and Orson* allows Orson to enter society. The removal of the right to speak is a removal of freedom and self-government which are, for Overton, essential to the establishment of a true humanity. The inability to speak might also be related to the desire for the laws to be available in English. French was the official legal language, and the availability of the law to the common man was central to the ideas of Edward Coke and the Levellers.[71] Coke's *Reports* were first printed in English in 1658.

The translation of the law is a reclaiming of the lost rights to liberty; unlike Bacon's naming, however, dominion is not Overton's aim. His aim is shared understanding. For David Joris, the mid-sixteenth-century Friesian Anabaptist, the restitution of the original order was to be achieved not through naming – which is to be translated (as it is in Bacon) as a form of dominion – but through conversation, a communal endeavour. According to a contemporary chronicler, Joris saw himself as 'the new God who led the simple folk to believe that he could speak with all tongues to the wild beasts and birds'. It was in his 'frequent references to the "new Adam" who would regain Edenic innocence' that Joris proposed his own version of the Baptist equality.[72] Overton never claims that he can speak to animals, but there is a similar sense of the possibility of community in his work.

In the extract from *A Defiance* Coke emerges again, more significantly, when the existing proofs of human-ness which Overton notes – wealth, stature and shape – are examined. Two of these are clearly related to Coke's discussion of inheritance. His notion of human status is revealed through possession of property which in turn is reliant on shape. But Overton seems to be aware of the problems and limitations of this idea: it is formulated within a fallen world; reclamation represents the attempt to return to perfection, it can never in itself represent perfection. This is why wealth and shape are all that is left to represent humanity in the current political climate.

The issue of stature returns us to the emblems discussed above in Chapter 1. The dwarf replaced the ape in George Wither's *Emblems*, and this replacement brings into play the significance of height in the delineation of human-ness. The dwarf is an '*Apish-Pigmie*' and is removed from humanity.[73] This removal is noted by Overton in *A Defiance*, and is rectified by him on two occasions. In *Mans Mortallitie* he writes:

> a *Giant* is no more a Man then a *Dwarfe*: there may be a graduall distinction, and yet no *Essentiall* defference; Degrees of *Faculties* in severall persons, and yet the *Faculties* the same, and of one nature, though not equally excellent: and the Degree doth not make a *Facultie* more a *Facultie*, or lesse a *Facultie* ...[74]

This is repeated in *An Appeale From the degenerate Representative Body* in 1647, where he writes:

> a Dwarfe is as much a man as a Gyant, though not so bigge a man; and so, though the gifts and graces of God are one *radically*, yet different in their *species*, and all from one and the same spirit, which can act nothing contrary to its owne nature ...[75]

Stature counts for nothing in the establishment of human-ness in Overton's thought. Wither's replacement for the ape is reintroduced into human society and can, implicitly, possess in Coke's designation: the dwarf has human shape. There is no evidence that Overton read Wither's emblem, but it seems significant and appropriate that the dwarf re-emerges here. The emblematic collapse of the boundary between human and animal was to be rectified in the political order.

But if a direct link cannot be made between Overton and Wither's emblem Wither's translation of Nemesius' *Of the Nature of Man* was a central source in *Mans Mortallitie*. In his Preface to the translation, Wither writes:

> By the first *sinne*, wee lost, indeed, our *light*, but not our *eyes*. And therefore, when god sent the *light* of mankind into the World, hee reprobated those, onely, who *rejected* it. And why? not because *they saw it not*; but *because they loved it not*.[76]

This is a return to the Baptist principles discussed earlier in the chapter. It is in the rejection of Christ – in a human act – that reprobation occurs. The Jews, Wither argues, saw Christ and still rejected him, and for this they are truly reprobate.[77] The exclusion of the Jews is one which Overton questions, but the reintroduction of the dwarf is not merely important because of the further link with Wither which it might offer, it is also significant because it offers a way of understanding the issues raised by the apparently exclusive franchise which is at the heart of Leveller ideas. For Overton, at least, it is in the innate and God-given status of humanity that true freedom exists, and this is where we need to relocate the exclusions which he, with the other Leveller leaders, proposed.

IV

Wives and servants are excluded from the Leveller proposals for the franchise because, Brian Manning argues, they 'could not be expected to vote any other way than their master required them'. Those in receipt of alms 'were the responsibility of the community … and had the same relationship of dependence as wives, children, and servants to the family; so that it could be conceived that the community cast the votes on their behalf just as the father voted on behalf of his family.' He proposes that the other group of beggars – the voluntary unemployed – had no rights to vote because the Levellers 'derived rights from labour'.[78] This is a point which Manning takes from Macpherson, who implies that a distinction between master and servant can be made on the basis of whether an individual has alienated themselves through work. Macpherson writes:

> The economic rights which were not constitutionally guaranteed but were left to the good sense of future parliaments, could in any case be best secured by those who had a direct personal interest in them, namely, *those who had not alienated the use or direction of their own energies and capacities by becoming wage-earners or recipients of alms.*[79]

Migration to the city created 'feelings of alienation'; what Macpherson seems to imply is that these feeling were self-made.[80]

Such an emphasis on human agency denies the economic imperatives which may have led people to abandon an established rural or provincial domestic environment.[81] But Macpherson's interpretation also fails to recognise the sense of inevitability in Overton's thought. Beneath the rhetoric of choice which would allow for self-alienation, the refusal of self-propriety as a parallel of the refusal of Christ, is a broader vision which regards humans as always-already fallen and as living in an imperfect world. The status of 'servant', as Keith Thomas notes, is more limited than Macpherson allows for, but it is also to be interpreted as far more general: it represents a loss of self-propriety which is a result of the general corruption of the world.[82]

The economic system which creates servants and the political situation in which there are exclusions based on labour and status are a reiteration of the postlapsarian biblical situation. After the Fall Adam is told, 'In the sweat of thy face shalt thou eat bread' (Genesis 3: 19); labour represents his alienation from animals, but more importantly from God and perfect order. In the same way it is labour which removes humans from freedom. To earn wages is to be removed from the franchise; to have a say in government is to have certain rights in postlapsarian society. The limited nature of the Leveller franchise is a result of the corruption of the world. In the postlapsarian order it is only in death and resurrection that true and fair government can be found. In the changed order, however, there would be no unnatural dominion over beasts or over humans. There are no servants in the new order; there is a true freedom for all on earth as in heaven.

The link between the Fall and the corruption of government is important here. The return to an original order of government has been recognised as a desire to throw off the 'Norman yoke' and return to the pre-1066 social order.[83] This is true for Overton as for the other Levellers, but what is also to be traced in Overton's thought is a return to a prelapsarian perfection.[84] Where Christopher Hill notes the movement in Leveller thought from 'historical mythology to political philosophy' I am arguing that, for Overton at least, the movement was, in a very specific way, from theology to political philosophy; from mortalism to liberty.[85] The Fall narrative which is offered as the scriptural origin of animal wildness and human domination of the natural world is also interpreted as the scriptural origin of human bestialisation and exploitation. Political struggle and political change are not only national issues, they are natural ones. Because of the Fall there was a descent into human tyranny and dominion, and this created a political system in which the natural order has broken down. It is in this emphasis on religious ideas that there is a possibility of a return to perfection, or 'achieved self-hood', within Overton's ideas.[86] Human status is never lost forever, and, a change in the political order would allow for true human-ness to be achieved.

This link between divine and earthly government can be traced more generally in mid-seventeenth-century ideas of freedom. The order of government which was proposed was tied up with a belief in the ultimate government of God. 'The movement is not simply one from being under authority to being free from authority. Freedom from inferior or inappropriate authority is preliminary to subjection to a higher and more appropriate *and therefore more legitimately demanding* authority.'[87] If government is resistable it is implicitly not natural: natural government is the government of God. 'Full humanity', J. C. Davis notes, is not to be sought in liberation or freedom itself but in 'instrumentality, in accepting and submitting to a preordained but varying role under God's will'.[88]

Mans Mortallitie, a text to which Davis makes no reference, seems in this context archetypal. In acknowledging human instrumentality as evidence of animal status Overton not only uses the animal in the same way that Perkins, Draxe, Rawlinson and Goodman before him used it – as a tool of humans – he also recognises the impossibility of ever truly separating human from beast. Human-ness is achieved through an acknowledgement of animality: this is not anthropomorphism or misanthropy but is a coming together of religion, humanism, science and the law. The exclusivity of the Leveller proposals for the franchise is, along with Coke's designation of the human on the basis of wealth and shape, an acknowledgement of the inevitable corruption of human society. Instrumentality is a freedom when God is the ruler, but in the face of abitrary rulers it is, like instrumentality in the law, the role of the animal.

What is missing from humanity for Overton – what the emphasis on wealth, stature and shape make up for – is the political climate in which '*that Birth-right of all English men*' – freedom – is to be found.[89] Where Coke argues that the loss of access to the common law inhibits inheritance, in *A Defiance* Overton emphasises the ultimate loss which comes with the loss of the birthright: the loss of human status. The human will always be removed from the animal by a potential and not actual distinction until the existing political order is overturned.

Humanity's natural and innate rights have been usurped and the one reason which Overton could cite as differentiating the species, having denied the soul, has been overturned. But Overton refuses to abandon the idea of a difference between the species, and what is significant in his refusal is his recognition that the human is never *a priori* but is always *a posteriori*. He sees the possibility of a distinct human status but also understands that this can be lost. The struggle to establish the always-already human which so upset the logic of previous formulations in Reformed ideas, humanism, the new science and the law is regarded as futile by Overton. All that can be sought is a status which accepts and recognises its created nature. The making of the human is a political struggle without end, but which must take place. It is with Overton's lost and found humanity that I conclude this chapter.

V

In *The Arraignment of Mr Persecution*, cited earlier, Overton offers the natural world as a figure of communal living: 'the *Oxe*, the *Asse*, the *Sheep*, and all sorts of *Cattell*; the *Dove*, the *Sparrow*, all kinds of *Birds* have a State harmony, a publike Toleration, generall Concord and unity among themselves in their severall kinds'.[90] In this image Overton allows for difference – of several kinds – and community, and this is a notion of coexistence which he also places at the heart of his ideal human community. In *An Appeale From the degenerate Representative Body* the operations of this communal living are spelt out:

it is an *Ordinance amongst men* and for *men*, that *all men* may have *an humane* subsistence and safety to live as *men* amongst *men*, none to be excepted from this human subsistance, but the unnaturall and the inhumane, it is not for *this opinion*, or that *faction*, this Sect or that *sort*, but equally and alike indifferent for all men that are not degenerated from humanity and humane civility in their *living* and *neighbourhood*: And therefore the destroyers and subverters of humane society, safety, cohabitation and being, are to be corrected, expulsed, or cut off for the preservation of safety, and prevention of ruine to both *publike* and *private*: and thus is *Magistracy* for the praise of them that doe well, and for the punishment of those that doe evill.[91]

This passage proposes a clear and unequivocal distinction of human from inhuman based on the ability to live communally, to be tolerant of difference. The dangers of the loss of human-ness are represented in the same way as Coke saw the dangers of the loss of access to the common law: safety is at stake. The safety for Overton, I suggest, is species safety. Being human is existing in a human way: as in the Pelagian ideas which can be traced in Overton's religious beliefs, it is in actions that status is achieved. Humans can choose to be inhuman, a status very different from non-human or animal, but if such a choice is made then natural government – the government of God – is in place to punish.

Overton recognises the possibility of degeneration – the fact that the things which separate the human from the animal like reason, consideration and science, must be augmented by learning and can be diminished through negligence. In recognising this he has to abandon any sense of human dominion, for such dominion would destroy all notions of tolerance which are so central to his ideas (as shown in the guilt of Thomas Edwards, Mr Persecution himself). The animal, in Overton's thought, is both different and same, and it is in the acknowledgement of this paradox that a new notion of humanity emerges.

All separations between the species are both precursors to and representations of other divisions. The birthright of humanity is not tyranny in any of its forms (natural or political) but liberty, and this liberty, as the Pelagian position would imply, includes the liberty of choice. But, again, as a parallel to the Pelagian position, the wrong choice can destroy status: Hell 'is but in *poße*, not in *eße* till the Resurrection.'[92] It is humans who make their own hell, and it is humans who make their own paradise. The animals do not have this choice, not because they are incapable of choice, but because they have never in themselves made the wrong choice. Animals are innocent of the Fall and for this reason they are saved; by extension the masses are innocent of the corruptions of power and for this reason they deserve, and can achieve, liberty. Human and animal alike, their sufferings are the result of the corruption of their divinely instated steward.

By offering an image of an inclusive and yet indistinguishable humanity Overton underlines the fundamental need for social change. Toleration is important – all humans; magistrates, savages, women, dwarves and giants, are included – but the true expression of human-ness can only be achieved through the overturning of the political order and the return to a paradisal state of self-propriety; to a time when wildness – the anti-social and uncivil – was not part of the order of things.

The representation of the animal in theological texts by writers such as Perkins, Goodman, Herbert and Draxe finds its parallel in the figure of the human in Overton's work. Just as the animal is included in religious ideas but excluded from the church in Goodman's work, so the disempowered human is included in society but excluded from the institutions of power in Overton's. These exclusions mean that humans lose their birthright and there-fore their humanity, just as animals, for Goodman, lose the afterlife and their place in God's kingdom. Overton's recognition of the anthropocentrism of orthodox theology is just part of his return to the ideal. The destruction of humanity which can also be traced in humanism, the new science and the law is corrected in Overton's ideas. The bestialisation of the understanding, the futility of naming and the inaccessibility of the institution of law and its benefits are redressed through political change which reinstates judgement and proposes communal understanding and shared access to power. Political change would return humanity to their status as humans, and would be paralleled by the reintroduction of a natural order based on conversation rather than naming. Overton's animals are not named among the beasts and certainly do not say 'Eat me'; they are calling for revolution.

Epilogue: Return to the Bear Garden

On 2 July 1649, less than six months after the execution of Charles I, Richard Overton 'staged' a metaphorical baiting in which the Leveller dogs took on the great Bull of Bashan.

> Whu – all my brave Levelling Bull dogs and Bear dogs, where are you? Siz – , ha – looe – ha – looe – all fly at him at once: There at him, at him; O brave Jockey with the Sea-green ribbond in his eare! that Dog and his fellow for fourty shillings a Dog: Hold, hold, he hath caught him by the Gennitals, stave him off, give the Bull fair play. – A pox – they have burnt my Dogs mouth.[1]

Overton's Defyance of the Act of Pardon was written from the Tower of London and aimed to assure his followers that 'my silence hath not proceeded from any degeneration or instability in me to that *Righteous Cause*'.[2] It did not have quite the desired effect.

The source for the baiting was Psalm 22: 12–13: 'Many Bulls have compassed me: strong bulls of Bashan have beset me round. / They gaped upon me with their mouths, as a ravening and a roaring lion.' Overton placed himself and his followers in the position of David, and Cromwell became the figure of the despair they felt. The alignment of Cromwell with the bull was horribly appropriate because the bull is like a lion, and the lion, as emblematic literature has always noted, is like a King. Cromwell may have helped to rid the country of Charles I, but he was in the process of replacing the King with another large and violent animal, himself. The baiting of the bull of Bashan is an attack on all that keeps David from God, on all that keeps the Levellers from power.

Exactly a week later, in response to the hostile reception of this text, Overton was forced to explain himself.

> All pallates are not pleased with that sheet intituled *Overtons Defiance, &c.* yet falleth it out no otherwise then I expected; It seems many are weak and as many are offended, and chiefly with that figurative passage of the *Bull*; especially at the word *Pox* ...[3]

167

Through the baiting Overton had attempted to defend his 'integrity', but his language offended his readers.[4] His use of 'uncivill' terms is excused by citing scriptural precedent: '*Eliah* mocked the Priests of *Baal*, and said, *Cry aloud* ... Sure this was a jear to some purpose ... then why may not I cry – ha – looe – ha – looe – &c. and not be condemned?'[5]

The distaste of his readers, however, does not find Overton unprepared: 'yet falleth it out no otherwise then I expected'. He argues that they 'need not' take offence

> did they but also take into their thoughts, *the adulterous and wicked generation*, on whom that Metaphor is made good, *a people whose heart is waxed grosse, and their ears dull of hearing, having closed their eyes, least at any time they should see, hear, understand and be converted.*[6]

Overton invokes Paul preaching to the Jews in Rome (Acts 28: 27) and aligns himself with the bringer of the Gospel. If the readers had realised that 'The figure is but the shell' they would have 'crack[ed] the shell to take out the kernell', made the metaphor good, and revealed themselves to be truly converted. As it is they did not, and by taking offence they reveal their misunderstanding not just of Overton himself but of the cause for which he has been imprisoned. They are, as Sir Philip Sidney wrote, 'well worthy to have [their names] chronicled among the beasts he writeth of'; not humans but dogs.[7] Just as human can never be truly separated from beast, so Leveller can never be a stable and long-lasting category. Overton then writes 'you ... trample [the meaning] under your feet, turne againe, and are (some of you) ready to rent me; He that had cast Pearls before Swine could have expected no lesse.'[8] He voices the decline of the cause: his followers have lost the ability to interpret, become like the Jews who deny the truth of Christ, and have turned on him. Overton's representation of himself slides from David and Paul to the ultimate martyr, Christ.

In his use of baiting as a metaphor for their political struggle Overton follows Dekker's alignment of the oppressed with the animal, uses the religious parable as a means of expressing a moral and invokes the humanist link between interpretation and humanness. But he also acknowledges that cruelty will never end without political change.

But stay, first let me clap this nettle under his Tayle, and tell him wee'l never leave bating, if all the lusty levelling Masties in *England* will do it, till we have worryed, or broke the Buls neck, or else gain'd our Agreement.[9]

The struggle between human and animal which is played out in the Bear Garden is not merely a figure of political struggle: it is the same as the political struggle. Both are the result of a larger battle which was set in motion by the Fall. However, by acknowledging the animality of humanity Overton has prepared the ground for the distaste of his readers: if humans are animalised by the political situation then it is not surprising that those same humans will interpret like animals and misunderstand his meaning. In recognising the seriousness of the fight, in taking it to natural as well as political heights, Overton also reveals how difficult the fight will be. In fact he himself shares the difficulty of his supporters.

Overton writes, 'I confesse I love the sport.'[10] Even as he declares the impossibility of separating humans from animals in the fallen world Overton loves the Bear Garden, and in his declaration of love he disturbs the sense of equality which can be traced in his earlier work. A theory is only a theory, as Jonson noted of Sidney's humanism; what is needed is practice. Confessing to loving baiting reveals that the 'degeneration or instability' which Overton denied is firmly in place. Political power is still degenerate even though it has changed hands, and Overton himself remains unstable because of the fallen nature of power. In his love of baiting Overton recognises the viciousness of the animals and the cruelty of the spectators, and is trapped as both beast and spectator in the Bear Garden: he is both perceiving animals and a perceiving animal.

I am disappointed that Overton is a lover of bear-baiting. It would make the ending of this book much easier if I could go back to the Bear Garden to show its destruction; if I could offer a reading of an anti-baiting pamphlet by Overton – but he doesn't write one, so I can't. All I can offer is Richard Overton, a man whose work seems to draw together so many of the ideas which I found in my search to understand baiting, fighting to find a way of maintaining a struggle to change the fallen world in which political rule is arbitrary and animals are exploited.

But the struggle looks to be lost. In clinging on to his beliefs Overton is also clinging on to his human status, and he does this through the image of baiting. There is a terrible logic in this return to the Bear Garden. Baiting is the most spectacular representation of human dominion in early modern England, but it is also the most spectacular representation of humanity's failure to establish its human-ness. The anthropocentrism of *Overton's Defyance* is a step back from the call for change found in his earlier works. It is a retreat to the Bear Garden even as it proclaims its status, through the image of baiting, as a new attack. Changing the system of government was supposed to alter the order of nature, but the change in power which has taken place has not really changed things at all. If anything, things are worse: the 'Agreement of the People' ('my *all in all*') is no longer a possibility.[11] The acknowledgement of the love of baiting lays bare the Levellers' defeat.

Overton's Defyance and *The Baiting of the Great Bull of Bashan unfolded* were his final solo Leveller writings. After these two pamphlets all that can be attributed is the reissued, extended version of *Mans Mortallitie* in 1655. The struggle, and with it the dream of the new order, had gone. Overton turned away from this world and looked to the next to find an image of true community. At the end of *The Baiting of the Great Bull* he asked:

> If your neighbours Oxe, or his Asse were in a ditch, it is a shame to passe by and not to help; and behold, here's *all in the ditch*, then, why venture you not your time, your labours, your monies, &c. to redeem out all, *our Cause*, the nation, and us *in it*, and *with it*.[12]

The question was answered by Overton himself: we cannot see the truth because we perceive like animals.

Notes

Introduction

1. The magisterial study of this issue in the early modern period is Keith Thomas, *Man and the Natural World: Changing Attitudes in England 1500–1800* (London: Penguin, 1984). For important works on the human perception of animals in other periods see Joyce E. Salisbury, *The Beast Within: Animals in the Middle Ages* (London: Routledge, 1994); Harriet Ritvo, *The Animal Estate: The English and Other Creatures in the Victorian Age* (London: Penguin, 1990); and Kathleen Kete, *The Beast in the Boudoir: Petkeeping in Nineteenth-Century Paris* (London: University of California Press, 1994).
2. Stephen Greenblatt, *Shakespearean Negotiations: The Circulation of Social Energy in Renaissance England* (Oxford: Clarendon, 1988), p. 1.
3. Ibid., p. 20.
4. Keith Thomas notes that 'thirty or so copies of the Gutenberg Bible printed on vellum in 1456 used the skins of some five thousand calves'. Figures from Bühler, *The Fifteenth-Century Book*, cited in Thomas, *Man and the Natural World*, p. 25.
5. Thomas Taylor, *A Vindication of the Rights of Brutes* (1792), British Library classmark 722c49.
6. John Simons, 'The Longest Revolution: Cultural Studies after Speciesism', *Environmental Values*, 6 (1997), 491.
7. Ben Jonson, 'To Penshurst' (1612), in C. H. Herford and Percy Simpson, ed., *Ben Jonson* (Oxford: Clarendon, 1927), Volume VIII, p. 94.
8. Thomas Carew, 'To Saxham' (*c*.1640), in R. G. Howarth ed., *Minor Poets of the Seventeenth Century* (London: Dent, 1953), p. 86.
9. George Herbert, 'Providence' (1633), in C. A. Patrides, ed., *The English Poems of George Herbert* (London: Dent, 1974), p. 129.
10. Friedrich Nietzsche, *The Gay Science* (1882), translated by Walter Kaufmann (New York: Vintage, 1974), pp. 167–8.
11. Ibid., p. 167.
12. Francis Bacon, *Novum Organum* (1620), in James Spedding, Robert Leslie Ellis and Douglas Denon Heath, ed., *The Works of Francis Bacon* (1859; reprinted, Stuttgart: Friedrich Frommann, 1963), Volume IV, p. 54. I return to Bacon in more detail in Chapter 4. On the similarities of Nietzsche and Bacon, see Tony Davies, *Humanism* (London: Routledge, 1997), pp. 105–6.
13. Tho: Scot, *Philomythie or Philomythologie wherin Outlandish Birds, Beasts, and Fishes, are Taught to Speak true English plainely* (1616), 'Sarcasmos Mvndo, Or the Frontispiece explained', sigs. ¶ᵛ-¶2ʳ. Margot Heinemann notes the difficulty of the attribution of this work. There were two Thomas Scots writing in the early seventeenth

171

century, and Heinemann herself seems to align the two: both, she argues, were linked with 'opposition political circles abroad with which Pembroke sympathised'. Heinemann, *Puritanism and Theatre: Thomas Middleton and Opposition Drama under the Early Stuarts* (Cambridge: Cambridge University Press, 1980), pp. 113, 167 and 275–6.

14. Diana Fuss, 'Introduction: Human, all too Human', in Fuss, ed., *Human, All too Human* (London: Routledge, 1996), pp. 5–6.

15. Emmanuel Levinas, 'The Name of a Dog, or Natural Rights', in Levinas, *Difficult Freedom: Essays on Judaism*, translated by Séan Hand (London: Athlone Press, 1990), p. 153.

16. Ibid., p. 152: David Clark, 'On Being "The Last Kantian in Nazi Germany": Dwelling with Animals after Levinas', in Jennifer Ham and Matthew Senior, ed., *Animal Acts: Configuring the Human in Western History* (London: Routledge, 1997), p. 168.

17. Throughout the book I am following Alister E. McGrath's definition of the term 'Reformed' to mean broadly Calvinist ideas. McGrath, *Reformation Thought: An Introduction*, second edition (Oxford: Blackwell, 1993), p. 8.

18. Margaret T. Hodgen, *Early Anthropology in the Sixteenth and Seventeenth Centuries* (Philadelphia: University of Pennsylvania Press, 1964), pp. 214 and 363.

19. *OED*, second edition (Oxford: Clarendon, 1989), Volume XVI, pp. 155–7.

Chapter 1

1. Caroline Barron, Christopher Coleman and Clare Gobbi, ed., 'The London Journal of Alessandro Magno 1562', *The London Journal*, 9 (1983), 144.

2. Giles E. Dawson, 'London's Bull-baiting and Bear-baiting Arena in 1562', *Shakespeare Quarterly*, 15 (1964), 99.

3. On this issue see Ellen K. Levy and David E. Levy, 'Monkey in the Middle: Pre-Darwinian Evolutionary Thought and Artistic Creation', *Perspectives in Biology and Medicine*, 30 (1986), 95–106.

4. Edward Topsell, *The Historie of Foure-Footed Beastes* (1607), pp. 3–4.

5. In his account of the Bear Garden, discussed in more detail below, Thomas Dekker calls the dogs chasing the ape 'two footemen that ran on each side of his old *Apes* face'. Dekker, *Worke for Armourours: Or, The Peace is Broken* (1609), sig.B2^r.

6. On the place of the ape in art, see H. W. Janson, *Apes and Ape Lore in the Middle Ages and the Renaissance* (London: Warburg Institute, 1952) and Anat Feinberg, '"Like Demie Gods the Apes Began to Move": The Ape in the English Theatrical Tradition, 1580–1660', *Cahiers Elisabethains: Sur la Pre Renaissance et la Renaissance Anglaises*, 35 (1989), 1–13.

7. Homi Bhabha, 'Of Mimicry and Man: The Ambivalence of Colonial Discourse', *October*, 28 (1984), 128. The parallel noted here was

brought to my attention in the work of Stallybrass and White; see n. 9.

8. Marjorie Spiegel, *The Dreaded Comparison: Human and Animal Slavery* (London: Heretic Books, 1988), p. 37.
9. Peter Stallybrass and Allon White, *The Politics and Poetics of Transgression* (Ithaca: Cornell University Press, 1986), p. 41.
10. See, for example, George Wilson, *The Commendation of Cockes and Cock-fighting* (1607).
11. John Taylor, *Bull, Beare, and Horse* (1638), in *The Works of John Taylor The Water Poet Not Included in the Folio Volume of 1630* (Manchester: Spenser Society, 1876), Volume III, pp. 59 and 68.
12. Another danger to humans arose in the Bear Garden when the bear escaped from its chains. For records of the escapes see G. E. Bentley, *The Jacobean and Caroline Stage*, Volume IV, p. 211; C. L. Kingsford, 'Paris Garden and the Bear-Baiting', *Archaeologia*, 20 (1920), 162.
13. Phillip Stubbes, *The Anatomie of Abuses* (1583), sigs.Qv^v–Qvi^r.
14. William Perkins, *The Whole Treatise of the Cases of Conscience* (1596) in *The Workes of that Famovs and Worthy Minister of Christ in the Vniuersitie of Cambridge, Mr William Perkins*, Volume II (1617), p. 141.
15. Robert Bolton, *Some Generall Directions For a Comfortable Walking with God* (1625), pp. 155–6.
16. Coral Lansbury, *The Old Brown Dog: Women, Workers and Vivisection in Edwardian England* (London: University of Wisconsin Press, 1985), p. 32.
17. Gilbert Katherens, the builder of the Hope Theatre which housed both plays and baitings, notes that 'Beares and Bulls' were baited in the theatre. Katherens, cited in Bentley, *Jacobean and Caroline Stage*, Volume VI, p. 201.
18. Hunkes and Stone are two of the more famous bears. According to the list given by John Taylor in 1638, of the nineteen bears in the Bear Garden only one – Beefe of Ipswich – did not have a human name. Others included Ned of Canterbury, George of Cambridge, Blind Robin, Besse Hill and Kate of Kent. Taylor, *Bull, Beare, and Horse*, pp. 61–2.
19. E. P. Thompson, 'Patrician Society, Plebian Culture', *Journal of Social History*, 7 (1974), 389.
20. As Umberto Eco argues, carnival is never truly rebellious, it 'can only exist as an *authorised* transgression'. Eco, 'The frames of comic freedom', in Thomas A. Sebeok, ed., *Carnival!* (Berlin: Mouton, 1984), p. 6.
21. Spiegel, *Dreaded Comparison*, pp. 82 and 84.
22. Les Brown, *Cruelty to Animals: The Moral Debt* (London: Macmillan, 1988), p. 3. In this I am disagreeing with Keith Thomas who argues that in the early modern period what existed was 'the cruelty of indifference. For most persons, the beasts were outside the terms of moral reference.' My argument is that if there is pleasure in cruelty, which there obviously is in the Bear Garden, then that pleasure must in some ways be based upon the knowledge of the suffering of animals. Thomas, *Man and the Natural World*, p. 148.

23.   Raphael Holinshed, *Chronicles of England, Scotland and Ireland* (1587), Volume IV, p. 895.

24.   Dekker, *Worke for Armourours*, sigs.B1^v–B2^r.

25.   Ibid., sig.B2^r.

26.   Donald Lupton, *London & The Country Carbonadoed and Quartered into Severall Characters* (1632), p. 66.

27.   On the issue of civilisation in the wilderness see John Block Friedman, *The Monstrous Races in Medieval Art and Thought* (London: Harvard University Press, 1981), p. 148; Hayden White, 'The Forms of Wildness: Archaeology of an Idea', in Edward Dudley and Maximilliam E. Novak, ed., *The Wild Man Within: An Image in Western Thought from the Renaissance to Romanticism* (London: University of Pittsburg Press, 1972), pp. 3–38.

28.   On Aristotle, see Anthony Pagden, *The Fall of Natural Man: The American Indian and the Origins of Comparative Ethnology* (Cambridge: Cambridge University Press, 1982), pp. 43–4. On the development of the ideas of racial difference in early modern thought see Margaret T. Hodgen, *Early Anthropology in the Sixteenth and Seventeenth Centuries* (Philadelphia: University of Pennsylvania Press, 1964), and J. S. Slotkin, ed., *Readings in Early Anthropology* (Chicago: Aldine, 1965).

29.   Stanley L. Robe, 'Wild Man and Spain's Brave New World', in Dudley and Novak, ed., *The Wildman Within*, p. 47.

30.   Mark A. Mastromarino, 'Teaching Old Dogs New Tricks: The English Mastiff and the Anglo-American Experience', *The Historian*, 49 (1986), 25.

31.   Edmund Spenser, *A viewe of the presente state of Irelande* (1596), in Rudolf Gottfried, ed., *Spenser's Prose Works* (Baltimore: Johns Hopkins University Press, 1949), pp. 102 and 114. On the unspeakable nature of the Irish filthiness, see Chapter 5.

32.   Spenser, *A viewe of the presente state*, p. 210. On the issue of Ireland and wildness see Norah Carlin, 'Ireland and Natural Man in 1649', in Francis Barker *et al.*, ed., *Europe and Its Others* (Colchester: University of Essex Press, 1985), pp. 91–111.

33.   Both cited in Winthrop D. Jordan, *White Over Black: American Attitudes Toward the Negro, 1550–1812* (Chapel Hill: University of North Carolina Press, 1968), pp. 71 and 65. In 1788 it was not a felony to kill a slave in Virginia. See David Brion Davis, *The Problem of Slavery in Western Culture* (Ithaca: Cornell University Press, 1966), p. 58.

34.   Jordan, *White Over Black*, p. 80. For a similar description of monsters see Ambroise Paré, *Of Monsters and Prodigies* (1573), in *The Works of that Famous Chirurgeon Ambrose Parey* (1678), pp. 585–6. Monstrous births are discussed in more detail in Chapter 5.

35.   Cited in Marcia Vale, ed., *The Gentleman's Recreations: Accomplishments and Pastimes of the English Gentleman 1580–1630* (Cambridge: D. S. Brewer, 1977), p. 30.

36.   Pagden, *Fall of Natural Man*, pp. 43–4.

37.   Dorothy Leigh, *The Mothers Blessing* (1627), pp. 38–9 and 44.

38.   Topsell, *Historie of Foure-Footed Beastes*, p. 4.

39. Nicholas Breton, *The Court and the Country* (1618), in Alexander B. Gossart, ed., *The Works in Verse and Prose of Nicholas Breton* (New York: A.M.S. Press, 1966), Volume II, pp. 5, 6 and 14.

40. After 1572 entertainers were regarded as vagrants and subject to punishment unless they were patronised by a peer. The Queen's warrant meant little more than that the entertainer would be left alone and allowed to perform by the civic authories. See A. L. Beier, *Masterless Men: The Vagrancy Problem in England 1560–1660* (London: Methuen, 1985), pp. 96–9.

41. John Taylor, *Wit and Mirth* (1630), in W. Carew Hazlitt, ed., *Shakespeare Jest Books* (London: Willis and Southeran, 1864), Volume II, pp. 65–6.

42. A similar misunderstanding of the meaning of the warrant (this time on the part of a rural JP) can be found in Thomas Nashe's *Pierce Penniless* (1592) in 'The Tale of a Wise Justice':

> Amongst other choleric wise justices, he was one, that having a play presented before him and his township by Tarlton and the rest of his fellows, Her Majesty's Servants, and they were now entering into their first merriment, as they call it, the people began exceedingly to laugh when Tarlton first peeped out his head. Whereat the Justice, not a little moved, and seeing with his becks and nods he could not make them cease, he went with his staff and beat them round about unmercifully on the bare pates, in that they, being but farmers and poor country hinds, would presume to laugh at the Queen's Men, and make no more account of her cloth in his presence.

In Nashe, *The Fortunate Traveller and Other Works*, J. B. Steane, ed. (London Penguin, 1985), pp. 85–6.

43. Figures cited in A. L. Beier and Roger Finlay, 'Introduction: The Significance of the Metropolis', in Beier and Finlay, ed., *London: 1500–1700: The Making of the Metropolis* (London: Longman, 1986), p. 2, quotations, p. 20. On the issue of migration see Peter Clark and David Souden, ed., *Migration and Society in Early Modern England* (London: Hutchinson, 1987). In *The Economic and Philosophical Manuscripts* (1844) Marx aligns the alienation experienced by the industrial worker with the loss of 'species being [which] means that one man is alienated from another as each of them is alienated from the human essence.' Marx, 'Alienated Labour' in David McLellan, ed., *Karl Marx: Selected Writings* (Oxford: Oxford University Press, 1977), p. 83.

44. See Richard Holt, *Sport and the British: A Modern History* (Oxford: Clarendon, 1989), pp. 16–17. Oscar Brownstein has noted that by law a bull had to be baited before the meat could be eaten and that this might provide a link between the sport and civic festivities. Brownstein, 'The Popularity of Baiting before 1600: A Study in Social and Theatrical History', *Educational Theatre Journal*, 21 (1969), 241.

45. Dennis Brailsford, *Sport in Society: Elizabeth to Anne* (London: Routledge, 1969), p. 204.
46. See W. Symonds, 'Winterslow Church Reckonings, 1542–1660', *Wiltshire Archaeological Magazine*, 36 (1909–10), 42–3.
47. Peter Burke, 'Popular Culture in Seventeenth-Century London', *The London Journal*, 3: 2 (1977), 144–8, quotes 147–8.
48. See, for example, Richard Rawlidge and Thomas Dekker, discussed below, and Edward Hake, *Newes out of Powles Churchyarde* (1579), and Francis Lenton, *The Young Gallants Whirligigg* (1629).
49. Henry Farley, *St. Pavles-Chvrch Her Bill for the Parliament* (1621), sigs.E4^{r-v}.
50. Keith Wrightson, *English Society 1580–1680* (London: Hutchison, 1982), p. 51.
51. The role of the church was taken over in many ways by the inn: commerce replaced religion as an organising principle in the community. See Peter Clark, *The English Ale House: A Social History* (London: Longman, 1983).
52. See F. J. Fisher, 'London as an "Engine of Economic Growth"', in J. S. Bromiley and E. H. Kossman, ed., *London and the Netherlands Volume IV: Metropolis, Dominion and Province* (The Hague: Martinus Nijhoff, 1971), pp. 3–16; and Fisher, 'The Development of London as a Centre of Conspicuous Consumption in the Sixteenth and Seventeenth Centuries', *T.R.H.S.*, 4th Ser., 3 (1948), 37–50.
53. We know women attended baitings in the capital because of the death toll when the Bear Garden collapsed in 1583. See Stubbes, *Anatomie of Abuses*, sig.Qiv^v.
54. Felicity Heal notes that there was also a decline in hospitality in the early modern period because 'the rich were charged with indulging new forms of expenditure on building, clothes, and banquets, and with the pursuit of their pleasures in London rather than the countryside where they naturally belonged.' Heal, 'The Idea of Hospitality in Early Modern England', *Past and Present*, 102 (1984), 81. For a contemporary report on the decline of hospitality see William Vaughan, 'Of Hospitality', in *The Golden Grove* (1608).
55. Ben Jonson, *Bartholomew Fair* (1614) in, C. H. Herford and Percy Simpson, ed., *Ben Jonson* (Oxford: Clarendon, 1927), Volume VI, 'THE INDVCTION', lines 85–93.
56. John Marston, 'Satyre IX', *Here's a toy to mock an Ape indeed*, in *The Scourge of Villanie* (1598), in Arnold Davenport, ed., *The Poems of John Marston* (Liverpool: Liverpool University Press, 1961), pp. 160 and 161. Paris Garden is an alternative name for the Bear Garden.
57. John Dando and Harry Runt, *Maroccus Extaticus. Or, Bankes Bay Horse in a Trance* (1595), sig.B3^r.
58. Thomas Dekker, *The Seuen deadly Sinnes of London* (1606), 'The Induction to the Booke', sig.A2^v.
59. This dating is from Kenneth Muir's introduction to the Arden edition of *Macbeth* (London: Routledge, 1988), p. xvii.
60. Richard Rawlidge, *A Monster late found out and Discovered: or, the Scourging of Tiplers* (1628), p. 2.

61. Richard Johnson, *Looke on me London* (1613), sigs.B1^v–B2^r. The same potential for descent had been figured by Edward Topsell fourteen years earlier in *Times Lamentation* (1599), p. 385.
62. Thomas Adams, *The Gallants Burden* (1612), in *The Workes of Tho: Adams* (1630), p. 15.
63. Pagden, *Fall of Natural Man*, p. 17.
64. For an interesting structuralist account of the use of animals in understanding human society see Edmund Leach, 'Anthropological Aspects of Language: Animal Categories and Verbal Abuse', in Eric H. Lenneberg, ed., *New Directions in the Study of Language* (Cambridge, Mass: MIT Press, 1966), pp. 23–63. William Meredith Carroll argues that the terms beast, beastly, beastliness, brutebeast, brute and brutishness 'may be considered favourite words' in Elizabethan prose. Carroll, *Animal Conventions in English Renaissance Non-Religious Prose (1550–1600)* (New York: Bookman Associates, 1954), p. 46. On the historical context of the increasing use of these terms see Gary B. Nash, 'The Image of the Indian in the Southern Colonial Mind', in Dudley and Novak, ed., *Wild Man Within*, p. 71.
65. Breton, *Court and the Country*, p. 14.
66. Geffrey Whitney, *A Choice of Emblemes and Other Devices* (1586), p. 145.
67. Rosemary Freeman has called Whitney's *Emblemes* 'a storehouse for Elizabethan commonplaces'. Freeman, *English Emblem Books* (London: Chatto and Windus, 1948), p. 57.
68. Beryl Rowland, *Animals with Human Faces: A Guide to Animal Symbolism* (London: Allen & Unwin, 1974).
69. George Wither, *A Collection of Emblems Ancient and Modern* (1635), p. 14.
70. Microphilus, *The New-Yeeres Gift* (1636), sig.F5^v. See also E. W. Fairholt, *Remarkable and Eccentric Characters* (London: Richard Bentley, 1849), pp. 63–75. Hudson is mentioned in the broadsheet, *The Three Wonders of this Age* (1636). The other wonders are William Evans, the King's giant porter, and Thomas Parre, who was 103 years old. The broadsheet is attached to John Taylor, *The Old, Old, Very Old Man* (1635), British Library c34f43.
71. Thomas Fuller, *The History of the Worthies of England* (1662), (Reprinted, London, 1811), Volume II, pp. 243 and 244.
72. Wither, *Collection of Emblems*, p. 14.
73. [Thomas Morton], *A Treatise of the threefolde state of man* (1596), p. 34.

Chapter 2

1. William Perkins, *A Discourse of Conscience* (1596), in *The Workes of That Famovs and Worthy Minister of Christ in the Vniuersitie of Cambridge, Mr William Perkins* (1616–18), Volume I (1616), p. 517.
2. See R. T. Kendall, *Calvin and English Calvinism to 1649* (Oxford: Oxford University Press, 1979).
3. Ralph Houlbrooke, 'The Puritan Death-bed, c.1560–c.1660', in Christopher Durston and Jacqueline Eales, ed., *The Culture of English*

Puritanism, 1560–1700 (Basingstoke: Macmillan, 1996), p. 124. For the reasons outlined in Durston and Eales' useful introduction to this collection I avoid the term 'puritan' throughout the rest of this book. See Durston and Eales, 'Introduction: The Puritan Ethos, 1560–1700', especially pp. 1–6.

4. Ian Breward, 'Introduction' to Breward, ed., *The Work of William Perkins* (Appleforth: Sutton Courtenay Press, 1970), p. 42.
5. Giovanni Pico della Mirandola, *On The Dignity of Man* (1486), translated by Charles Glenn Wallis (Indianapolis: Bobbs-Merrill, 1965), p. 5.
6. John Calvin, *Institutes of the Christian Religion* (1559 edition), translated by Henry Beveridge (London: James Clarke & Co, 1949), Volume I, pp. 210–11.
7. Ibid., p. 214.
8. Alister E. McGrath, *Reformation Thought: An Introduction*, second edition (Oxford: Blackwell, 1993), p. 240.
9. Stevie Davies, 'Introduction', in Davies ed., *Renaissance Views of Man*, (Manchester: Manchester University Press, 1978), p. 15.
10. Calvin, *Institutes*, Volume I, p. 215.
11. William Perkins, *A Treatise of Vocations or Callings of Men, With the Sorts and Kinds of them and the right use thereof* (1603), in *The Workes*, Volume I, p. 750.
12. Calvin, *Institutes*, Volume I, p. 181.
13. Ibid., pp. 171 and 172. Calvin is here invoking Matthew 10: 29, 'Are not two sparrows sold for a farthing? and one of them shall not fall on the ground without your Father.'
14. [Thomas Morton], *A Treatise of the threefolde state of man* (1596), pp. 5–6.
15. Calvin, *Institutes*, Volume II, p. 74.
16. William Perkins, *The Foundation of the Christian Religion Gathered into Six Principles* (1591), in *The Workes*, Volume I, p. 1.
17. John Moore, *A Mappe of Mans Mortalitie* (1617), p. 9.
18. Gervase Babington, *Certaine Plaine, briefe, and comfortable Notes vpon euerie Chapter of Genesis* (1592), fol.5ᵛ.
19. [Morton], *Treatise of the threefolde state*, p. 31.
20. William Perkins, *An Exposition of the Symbole, Or Creed of the Apostles* (1595), in *The Workes*, Volume I, p. 151.
21. Thomas Draxe, *The Earnest of our Inheritance* (1613), p. 26.
22. John Calvin, *A Commentarie of John Caluine, Vpon the first booke of Moses called* Genesis (1578), p. 75.
23. On this see Arthur O. Lovejoy, *The Great Chain of Being: A Study of the History of an Idea* (London: Harvard University Press, 1936).
24. Edward Cooke, *Bartas Junior. Or, The Worlds Epitome, Man* (1631), p. 41.
25. Perkins, *Discourse of Conscience*, Volume I, p. 517.
26. Thomas Aquinas, *Summa Contra Gentiles*, translated by James F. Anderson (New York: Doubleday, 1956), p. 267.
27. Thomas Aquinas, *Summa Theologiae*, translated by R. J. Batten (London: Blackfriars, 1975), p. 91. For a more detailed discussion of

Aquinas' understanding of animals see Peter Drum, 'Aquinas and the Moral Status of Animals', *American Catholic Philosophical Quarterly*, 66: 4 (1992), 483–8.

28. John Rawlinson, *Mercy to a Beast. A Sermon Preached at Saint Maries Spittal in London on Tuesday in Easterweeke, 1612* (Oxford: 1612), p. 33.
29. Rawlinson, *Mercy to a Beast*, pp. 23–31 and 46.
30. Ibid., p. 7.
31. Ibid., p. 27.
32. Thomas Adams, *Meditations Vpon Some Part of the Creed*, in *The Workes of Tho: Adams* (1630), p. 1133.
33. For an overview of the literature dealing with the animal soul and the possibility of animal immortality see Peter Harrison, 'Animal Souls, Metempsychosis, and Theodicy in Seventeenth-Century English Thought', *Journal of the History of Philosophy*, 31: 4 (1993), 519–44.
34. Draxe, *Earnest of our Inheritance*, pp. 5, 6 and 10.
35. Robert V. Schnucker writes that 'the Catholic Church had argued that children were not capable of mortal sin until the age of seven'. Schnucker, 'Puritan attitudes towards childhood discipline, 1560–1634', in Valerie Fildes, ed., *Women as Mothers in Pre-Industrial England, Essays in Memory of Dorothy McLaren* (London: Routledge, 1990), p. 114.
36. Calvin, *Institutes*, Volume II, p. 518.
37. William Gouge, *Of Domestical Duties* (1634), p. 528. On this issue see Leah Sinanoglou Marcus, *Childhood and Cultural Despair: A Theme and Variations in Seventeenth-Century Literature* (Pittsburg: University of Pittsburg Press, 1978), especially pp. 47–8.
38. W. H. [William Hubbocke], *An Apologie of Infants In A Sermon* (1595), p. 10.
39. 'The Ministration of Baptism to be vsed in the Church', *The Booke of Common Prayer* (1564), n.p.
40. Perkins, *Foundation of the Christian Religion*, Volume I, p. 5.
41. The ideas emerge in Calvin's thought, but are first given the clear theological designation 'double justification' in the ideas of Martin Bucer. See McGrath, *Reformation Thought*, pp. 111–12.
42. *The Second Tome of Homilies* (1563), fol.146ᵛ.
43. G. W. Bromiley, *Baptism and the Anglican Reformers* (London: Lutterworth Press, 1953), p. 32; Jonathan D. Trigg, *Baptism in the Theology of Martin Luther* (Leiden: E. J. Brill, 1994). I am indebted to Trigg's account in the following paragraphs. For the ideas of other Reformed thinkers see McGrath, *Reformation Thought*, pp. 170–85.
44. Luther, *Luther's Works*, 5, p. 247; and Barth, *Church Dogmatics*, p. 169, cited in Trigg, *Baptism in the Theology of Martin Luther*, pp. 20 and 3.
45. Ibid., p. 24. It is the lack of scriptural authority – 'appointed place' – which caused the reduction in the number of sacraments in the Reformed church from the Catholic seven (baptism, the eucharist, penance, confirmation, marriage, ordination, and extreme unction) to just two (baptism and the eucharist). See McGrath, *Reformation Thought*, pp. 159–87.

46. Trigg, *Baptism in the Theology of Martin Luther*, pp. 23–5, quote, p. 24. On the vestiarian controversy, see M. M. Knappen, *Tudor Puritanism: A Chapter in the History of Idealism* (1939; reprinted London: Chicago University Press, 1970), pp. 187–216.
47. Kendall, *Calvin and English Calvinism*, p. 4, n. 3.
48. See John Foxe, *A Sermon Preached at the Christening of a Iew, at London* (1578), n.p. The printed text includes 'the confession of faith, which Nathaniel a Iewe borne, made before the Congregation…'.
49. Trigg, *Baptism in the Theology of Martin Luther*, pp. 76, 103 and 106.
50. See e.g. Philip Greven, *The Protestant Temperament: Patterns of Child Rearing, Religious Experience and The Self in Early America* (New York: Meridian, 1979); and John Morgan, *Godly Learning: Puritan Attitudes Towards Reason, Learning and Education, 1560–1640* (Cambridge: Cambridge University Press, 1986).
51. H. C. Porter argues that 'the only pastoral answer to the problem [of predestination] was to assume that all pious hearers of the word were elected; however Calvinist in the study, the preacher must be Arminian in the pulpit'. Porter, *Reformation and Reaction in Tudor Cambridge* (Cambridge: Cambridge University Press, 1958), p. 312.
52. Perkins, *Christian Oeconomy, Or A Short Survey of the Right Manner of Erecting and Ordering a Family, according to the Scriptures*, in *Workes*, Volume III (1618), p. 694.
53. John Dod and Robert Cleaver, *A Treatise or Exposition Vpon the Ten Commandments* (1603), f.7^r.
54. William Coster, '"From Fire and Water": The Responsibilities of Godparents in Early Modern England', in Diana Wood, ed., *The Church and Childhood* (Oxford: Blackwell, 1994), p. 305.
55. *The Booke of Common Prayer*, n.p.
56. Dorothy Leigh, *The Mothers Blessing* (1627), pp. 25–6.
57. Figure from Keith Wrightson, *English Society 1580–1680* (London: Unwin Hyman, 1982), p. 190.
58. See Ian Green, '"For Children in Yeeres and Children in Understanding": The Emergence of the English Catechism under Elizabeth and the Early Stuarts', *The Journal of Ecclesiastical History*, 37: 3 (1986), 397–425. Green states that 'over 350 different catechitical forms or works can be traced, the vast majority of which, it should be added, were of English origin and were published after 1570' (400).
59. Henry Holland, *The Historie of Adam, or the foure-fold state of Man* (1606), sig.Aii^v.
60. Perkins, *Foundation of the Christian Religion*, Volume I, sig.A2^r.
61. Ibid., p. 1.
62. Ibid., pp. 3–4.
63. [Hubbocke], *Apologie for Infants*, p. 24.
64. Perkins, *Discourse of Conscience*, Volume I, p. 517.
65. Sir Philip Sidney, *Astrophil and Stella* (1591), 1, line 14, in Katherine Duncan-Jones, ed., *Sir Philip Sidney: Selected Poems* (Oxford: Clarendon, 1988), p. 117. On the similar issue of erected wit and infected will in *The Defence of Poetry* (1579) Alan Sinfield argues that Sidney's position is between 'two stools of protestant thought',

Calvin's and Hooker's. Sinfield, 'The Cultural Politics of the *Defence of Poetry*', in Gary F. Waller and Michael D. Moore, ed., *Sir Philip Sidney and The Reinterpretation of Renaissance Culture* (London: Croom Helm, 1984), pp. 135–6.

66. Perkins, *Discourse of Conscience*, Volume I, p. 535.
67. Ibid., p. 517.
68. Kendall, *Calvin and English Calvinism*, p. 9.
69. Perkins, *Discourse of Conscience*, Volume I, p. 547.
70. Holland, *Historie of Adam*, sig.Hhiii^v.
71. On the issue of the problem of double-predestination, see John Stachniewski, *The Persecutory Imagination: English Puritanism and the Literature of Religious Despair* (Oxford: Clarendon, 1991).
72. Calvin, *Institutes*, Volume II, p. 243.
73. William Perkins, *A Treatise Tending unto a Declaration Whether a Man be in the Estate of Damnation or in the Estate of Grace* (1586), in *The Workes*, Volume I (1616), p. 355.
74. Perkins, *Foundations of the Christian Religion*, Volume I, p. 6.
75. Roy Porter, 'Introduction' to Porter, ed., *Rewriting the Self: Histories from the Renaissance to the Present* (London: Routledge, 1997), p. 3.
76. Perkins, *Treatise Tending unto a Declaration*, Volume I, p. 383.
77. Perkins, *Discourse of Conscience*, Volume I, p. 517.
78. Ibid., p. 549.
79. Ibid., p. 517.
80. Ibid., p. 518.
81. Thomas Beard, *The Theatre of Gods Iudgements* (1597), p. 78.
82. Ibid., p. 139.
83. Ibid., p. 148.
84. Perkins, *Christian Oeconomy*, Volume III, p. 670.
85. Perkins, *Treatise Tending unto a Declaration*, Volume I, p. 384.
86. Perkins, *Discourse on Conscience*, p. 67.
87. Adam Douglas, *The Beast Within: Man, Myths and Werewolves* (London: Orion, 1992), pp. 91–2. The last wolf in England was in captivity in the Royal Menagerie in the Tower of London, 'kept on purpose because no wolves are to be found in England'. Gottfried von Bülow, ed., 'Diary of the Journey of Philip Julius, Duke of Stettin-Pomerania, through England in the Year 1602', *T.R.H.S.*, NS 6 (1892), 7.
88. George Gifford, *A Dialogue concerning Witches and Witchcraftes* (1593), sig.K2^v.
89. For a discussion of the case and other printed material surrounding it see Caroline Oates, 'Metamorphosis and Lycanthropy in Franche-Comté, 1521–1643', in Michel Feher with Ramona Nadoff and Nadia Tazi, ed., *Fragments for a History of the Human Body: Part One* (New York: Zone, 1989), pp. 315–16.
90. *A true Discourse Declaring the damnable life and death of one* Stubbe Peeter (1590), p. 7.
91. Ibid., p. 17.
92. Ibid., p. 16. Michael MacDonald has noted that the most serious crime at the time was a crime within the family: infanticide,

matricide, parricide. These crimes, he argues, were often understood in terms of madness. MacDonald, *Mystical Bedlam: Madness, Anxiety and Healing in Seventeenth-Century England* (Cambridge: Cambridge University Press, 1981), p. 128.

93. Perkins, *Discourse of Conscience*, Volume I, p. 517.
94. Richard Rawlidge, *A Monster late found out and Discovered: or, the Scourging of Tiplers* (1628), p. 28.
95. John Bulwer, *Anthropometamorphosis: Man Transform'd; Or, The Artificial Changeling* (1653), p. 521.
96. Oates, 'Metamorphosis and Lycanthropy', p. 317.
97. Bulwer, *Anthropometamorphosis*, p. 519.
98. Henri Boguet, *An Examen of Witches* (1590), translated by E. Allen Ashwin (London: Richard Clay & Sons, 1929), pp. 143–4.
99. Oates, 'Metamorphosis and Lycanthropy', p. 319.
100. Reginald Scot, *The discouerie of witchcraft* (1584), p. 97.
101. Ibid., p. 92.
102. Ibid., p. 93.
103. Douglas, *The Beast Within*, pp. 146–7.
104. Henry Holland, *A Treatise Against Witchcraft* (Cambridge, 1590), sig.F3^r.
105. James VI and I, *Daemonologie* (1603), pp. 60–1.
106. John Deacon and John Walker, *Dialogicall Discourses of Spirits and Devils* (1601), cited in MacDonald, *Mystical Bedlam*, p. 207.
107. John Webster, *The Duchess of Malfi* (1613), in Jonathan Dollimore and Alan Sinfield, ed., *The Selected Plays of John Webster* (Cambridge: Cambridge University Press, 1983), especially V. ii. 1–80.
108. Democritus Junior [Robert Burton], *The Anatomy of Melancholy* (1621; reprinted 1624), p. 8.
109. 'Almost all of the accused were of the lowest social status – vagrants, beggars, shepherds and peasants, some of whom were not natives of Franche-Comté; and as of 1598, many were women.' Oates, 'Metamorphosis and Lycanthropy', p. 326. These groups accused of transformation are parallelled in Pierre La Loyer's groups of '*Superstitious persons*': 'by reason of the imbecillitie of their nature .... we see that women and old men, are more addicted to superstition then any other'. La Loyer, *A Treatise of Specters or straunge Sights, Visions and Apparitions appearing sensibly vnto men* (1605), fol.104^v.
110. Lucien Malson, *Wolf Children*, translated by Edmund Fawcett, Peter Ayrton and Joan White, and published with Jean Itard, *The Wild Boy of Aveyron* (London: NLB, 1972), pp. 80 and 50.
111. On Digby's support of Decartes' ideas see Marjorie Nicolson, 'The Early Stage of Cartesianism in England', *Studies in Philology*, 26 (1929), especially 357–8.
112. Sir Kenelm Digby, *Two Treatises: In the One of Which, The Nature of Bodies; in the Other The Nature of Man's Soule* (1644), pp. 247–8.
113. Digby, *Two Treatises*, p. 248.
114. René Descartes, *Discourse on the Method* (1637), translated by F. E. Sutcliffe (Harmondsworth: Penguin, 1977), p. 74.

115. Recording the story of the Lithuanian wolf child, Bernard Connor writes, 'at length, being taught to stand upright, by clapping his Body against a Wall, and holding after the manner that Dogs are taught to beg; and being by little and little accustom'd to eat at Table, he after some time became indifferently tame.' Connor, *The History of Poland in Several Letters to Persons of Quality* (1698), pp. 342–3; see also pp. 348–9.

116. See, for example, Richard Bernheimer, *Wild Men In The Middle Ages: A Study in Art, Sentiment and Demonology* (Cambridge, Mass.: Harvard University Press, 1952), p. 85; Norah Carlin, 'Ireland and Natural Man in 1649', in Francis Barker *et al.*, ed., *Europe and Its Others* (Colchester: University of Essex Press, 1985), pp. 91–111; and John Block Friedman, *The Monstrous Races in Medieval Art and Thought* (London: Harvard University Press, 1981), p. 197.

117. See Hayden White, 'The Forms of Wildness: Archaeology of an Idea', in Edward Dudley and Maximillian E. Novak, ed., *The Wildman Within: An Image in Western Thought from the Renaissance to Romanticism* (London: University of Pittsburgh Press, 1972), p. 7.

118. Arthur Dickson, 'Introduction', in Dickson, ed., *Valentine and Orson* (London: EETS O.S. 204, 1937).

119. Throughout the following discussion I will refer to the tale as *Valentine and Orson*, although the two different versions have slightly different titles which will be used in endnotes: *The Hystory of the two Valyaunt Brethren* (1565), and *Valentine and Orson. The Two Sonnes Of the Emperour of Greece* (1637). There is also a fragment of the *c.*1505 text in the British Library which closely resembles the 1565 edition. This is catalogued as *The History of Valentine and Orson*.

120. These and other references are in *Hystory of the two Valyaunt Brethren*, n.p.

121. Ibid.

122. *Valentine and Orson*, p. 61.

123. Ibid., pp. 65–6.

124. Ibid., pp. 105–6.

125. Ibid., p. 125.

126. Lloyd De Mause, 'The Evolution of Childhood', in De Mause, ed., *The History of Childhood: The Untold Story of Child Abuse*, second edition (New York: Peter Bedrick Books, 1988), p. 31.

127. *Hystory of the two Valyaunt Brethren*, n.p.

128. *Valentine and Orson*, p. 232.

129. Pico della Mirandola, *Dignity of Man*, p. 6.

130. W. G. Craven, *Giovanni Pico Della Mirandola*, cited in Tony Davies, *Humanism* (London: Routledge, 1997), p. 95.

Chapter 3

1. Socrates quoted in Joanna Martindale, *English Humanism: Wyatt to Cowley* (Beckenham: Croom Helm, 1985), p. 32, Erasmus and Jonson cited by Martindale, p. 49, n. 72.

2. William Rankins, *A Mirrour of Monsters* (1587), f.17^r.

3. [Thomas Morton], *A Treatise of the threefolde state of man* (1596), p. 215.

4. William Perkins, *An Exposition of the Symbole, or Creed of the Apostles* (1595), in *The Workes of that Famous and Worthy Minister of Christ in the Vniuersitie of Cambridge, Mr William Perkins* (1616–18), Volume I (1616), p. 153.

5. Giovanni Pico della Mirandola, *On The Dignity of Man*, translated by Charles Glenn Wallis (Indianapolis: Bobbs-Merrill, 1965), p. 6.

6. Phillip Stubbes, *The Anatomie of Abuses* (1583), sig.Cv^v.

7. Roger Crab, *Dagons Down-fall; or, the great IDOL digged up Root and Branch* (1657), pp. 12–13. On Roger Crab, see Christopher Hill, 'The Mad Hatter', in *Puritanism and Revolution: Studies in Interpretation of the English Revolution of the Seventeenth Century* (1958; reprinted London: Penguin, 1990), pp. 303–10.

8. Adam Hill, *The Crie of England. A Sermon Preached at Paules Crosse* (1595), p. 38; Stubbes, *Anatomie of Abuses*, sig.Fviii^r.

9. Robert Crowley, *A Briefe Discourse against the outwards apparell and Ministring garmentes of the popishe church* (1566), sig.A4^r.

10. Stubbes, *Anatomie of Abuses*, sigs.Cii^v, Bviii^r and Ci^v.

11. Ibid., sigs.Cii^r and Nv^r.

12. John Rainoldes, *Th'Overthrow of Stage Playes* (1599; 1629 edition), p. 11. In a deposition from 1682 David Brown was accused of bestiality with a dog: 'as a Dogg uses to have with a Bitch'. Public Record Office, ASSI 45/13/2, 20. I look in more detail at depositions dealing with cases of bestiality in Chapter 5.

13. [Anthony Munday], *A Second and Third Blast of Retrait from Plaies and Theaters* (1580), pp. 95–6.

14. William Perkins, *A Discourse of Conscience* (1596), in *The Workes*, Volume I, p. 535.

15. See Tracey Hill, '"He hath changed his coppy": Anti-Theatrical Writing and the Turncoat Player', *Critical Survey*, 9: 3 (1997), 59–77.

16. This contradicts Michael O'Connell's reading of the important change from the very 'visual, sensual' worship of the Middle Ages – traced in pilgrimages, processions, the mystery cycles and so on – to the emphasis in Reformed thought on the Bible, the written word alone. O'Connell, 'The Idolatrous Eye: Iconoclasm, Anti-theatricalism, and the Image of the Elizabethan Theater', *ELH*, 52: 2 (1985), 288–9.

17. Crowley, *Briefe Discourse*, sig.B4^r.

18. William Perkins, *A Treatise Tending unto a Declaration Whether a Man be in the Estate of Damnation or in the Estate of Grace* (1586), in *The Workes*, Volume I, p. 384.

19. Despite this important difference between Reformed ideas and humanism there are many links to be made between the two philosophies. Humanist scholars helped to disseminate Luther's original attack on the established church in 1519; the return *ad fontes* – to the source – was at the heart of both movements. Alister McGrath, *The Intellectual Origins of the European Reformation* (Oxford: Blackwell, 1987), p. 64. For a more detailed discussion of the links between humanism and the Reformation see especially pp. 32–68.

20. Thomas North, 'To the Reader', in Anon., *The Morall Philosophie of Doni: drawne out of the ancient writers* (1570), (1601 edition), sig.A4^r.
21. North, 'Prologue' to *Morall Philosophie of Doni*, sig.B3^{r-v}.
22. *The Second Tome of Homilies* (1563), fol.153^r.
23. Roger Ascham's description of William Cecil's custom of hearing 'the minde of the meanest at his Table', has led Alan Stewart to write of the 'pretended levelling of humanism'. Stewart, *Close Readers: Humanism and Sodomy in Early Modern England* (Princeton, N.J.: Princeton University Press, 1997), p. 110.
24. Sister Mirian Joseph, *Shakespeare's Use of The Arts of Language* (New York: Columbia University Press, 1947), p. 10.
25. John Clarke, *Phraseologia puerilis Anglo-Latina* (1638), sig.A3^v.
26. Richard Mulcaster, *Positions* (1581), p. 5.
27. Ibid., p. 26.
28. William Gouge, *Of Domesticall Duties* (1634), p. 558.
29. Charles Hoole, *A New Discovery of the Old Art of Teaching School* (1660), in David Cressy, ed., *Education in Tudor and Stuart England* (London: Edward Arnold, 1975), p. 81.
30. Mulcaster, *Positions*, p. 27.
31. Perkins, *Discourse of Conscience*, Volume I, p. 551.
32. Annabel Patterson proposes that 'boys in the grammar schools would probably read Aesop as their first classical author.' Patterson, *Fables of Power: Aesopian Writing and Political History* (London: Duke University Press, 1991), p. 52.
33. [John Brinsley], *Esops Fables Translated both Grammatically, and also in propriety of our English phrase* (1624), sig.A5^r.
34. North, 'Prologue' and 'To the Reader' in *Morall Philosophie of Doni*, sigs.B1^v and A4^r.
35. North, 'To the Reader' in *Morall Philosophie of Doni*, sig.A4^r.
36. R. W. Maslen, *Elizabethan Fictions: Espionage, Counter-Espionage and the Duplicity of Fiction in Early Elizabethan Prose Narratives* (Oxford: Clarendon, 1997), p. 72.
37. John Moore, *A Mappe of Mans Mortalitie* (1617), p. 9.
38. George Puttenham, *The Arte of English Poesie* (1589), Gladys Doidge Willcock and Alice Walker, ed. (Cambridge: Cambridge University Press, 1936), p. 6.
39. E. K., 'Epistle' to Edmund Spenser, *The Shepheardes Calender* (1579), in J. C. Smith and E. de Selincourt, ed., *The Poetical Works of Edmund Spenser* (Oxford: Oxford University Press, 1924), pp. 416–19.
40. Sir Philip Sidney, *A Defence of Poetry* (1579), J. A. Van Dorsten, ed. (Oxford: Oxford University Press, 1989), pp. 62 and 64.
41. Both quoted in Richard Rambuss, *Spenser's Secret Career* (Cambridge: Cambridge University Press, 1993), p. 1.
42. Anon., *The Fables of Esope in English* (*c.*1570), sig.Fii^v. Another edition of the fables was printed in 1634. Unlike *Valentine and Orson*, apart from some modernising of the spelling, there are no changes between the two editions.
43. John Calvin, *Institutes of the Christian Religion* (1559 edition), translated by Henry Beveridge (London: James Clarke & Co: 1949), Volume II, p. 243.

44. North, 'The Prologue' to *Morall Philosophie of Doni*, sig.B3ʳ.
45. Henry Reynolds, *Mythomystes Wherein A Short Survay is Taken of The Natvre and Valve of Trve Poesy* (1632), sig.G1ᵛ. Reynolds is bewailing the corruption of modern poetry, arguing that it has lost the ability to speak the greater truths of the world.
46. Matthew 7: 6 reads, 'Give not that which is holy unto the dogs, neither cast ye your pearls before swine, lest they trample them under their feet, and turn again and rend you.' The text of *On The Dignity of Man* reads, 'But to disclose to the people the more secret mysteries, things hidden under the bark of the law and the rough covering of words, the secrets of the highest divinity, what was that other than to give what is holy to dogs and to cast pearls among swine.' Pico della Mirandola, *On The Dignity of Man*, p. 30.
47. Moore, *Mappe of Mans Mortalitie*, p. 43. The same verse of scripture is used by William Gouge in *Of Domesticall Duties*, p. 545. C. John Somerville has argued that to 'imagine the young as wild, undomesticated animals is evidence of a fundamental lack of affinity.' Sommerville, *The Discovery of Childhood in Puritan England* (London: University of Georgia Press, 1992), p. 86.
48. On the dangerous implications of the use of the term 'breaking-in', see Marjorie Spiegel, *The Dreaded Comparison: Human and Animal Slavery* (London: Heretic Books, 1988), p. 88.
49. Thomas Adams, *Lycanthropy, Or, The Wolfe Worrying the Lambs*, in *The Workes of Tho: Adams* (1630), p. 383.
50. Adams, *Lycanthropy*, p. 388.
51. In this I am disagreeing with Roger Chartier, who argues that the author has 'an interest in keeping close control over the production of meaning and in making sure that the text that they have written ... will be understood with no possible deviation from their prescriptive will.' Chartier, *The Order of Books: Readers, Authors and Libraries in Europe Between the Fourteenth and the Eighteenth Centuries*, translated by Lydia G. Cochrane (Cambridge: Polity Press, 1994), p. viii. In support of my reading David Norbrook suggests that Sidney's revision of *The Old Arcadia* was due to the problems this attempt at authorial liberality threw up: 'the first version ... was not didactic enough'. Norbrook, *Poetry and Politics in the English Renaissance* (London: Routledge and Keegan Paul, 1984), p. 104.
52. Ibid., p. 101.
53. Sir Philip Sidney, *The Old Arcadia*, Katherine Duncan-Jones, ed., (Oxford: World's Classics, 1985), p. 315.
54. Ibid., p. 315.
55. This summary is found in Alister McGrath's useful survey of some of the ways in which humanism has been interpreted. See McGrath, *Intellectual Origins*, pp. 32–3, quotation, p. 33.
56. Patterson, *Fables of Power*, p. 75.
57. Thomas N. Corns terms this 'the most elusive liberty to restrain'. Corns, 'The freedom of reader-response. Milton's *Of Reformation* and Lilburne's *The Christian Man's Triall*', in R. C. Richardson and G. M. Ridden, ed., *Freedom and the English Revolution: Essays in History and*

Literature (Manchester: Manchester University Press, 1986), p. 93. On Spenser's involvement in politics see Rambuss, *Spenser's Secret Career*.

58. Norbrook calls the place of Philisides' fable 'a crucial point' in the book. Norbrook, *Poetry and Politics*, p. 97.
59. Katherine Duncan-Jones 'Explanatory Notes', in Sidney, *Old Arcadia*, n. 64, p. 372; Alan Young, *Tudor and Jacobean Tournamments* (London: George Philip, 1987), p. 128.
60. Sidney, *Old Arcadia*, p. 225.
61. Patterson, *Fables of Power*, p. 70.
62. Norbrook, *Poetry and Politics*, p. 97.
63. Sir Philip Sidney, *A Discourse of Syr Ph. S. To The Queenes Majesty Touching Hir Mariage With Monsieur*, in Albert Feuillerat, ed., *The Prose Works of Sir Philip Sidney* (Cambridge: Cambridge University Press, 1963), Volume III, p. 52.
64. This interpretative strategy goes against William Dinsmore Briggs' unequivocal statement that 'it would be absurd to regard this fable as a poetical representation of the origin of man or of how he acquired dominion over the brutes'. Briggs, 'Political Ideas in Sidney's *Arcadia*', *Studies in Philology*, 28: 2 (1931), 152.
65. Sidney, *Old Arcadia*, p. 223.
66. Ibid., p. 224.
67. Pico della Mirandola, *On The Dignity of Man*, p. 6. Italics added.
68. Howard Needler, 'The Animal Fable Among Other Medieval Literary Genres', *New Literary History*, 22: 2 (1991), 423.
69. Annabel Patterson, *Censorship and Interpretation: The Conditions of Writing and Reading in Early Modern England* (London: University of Wisconsin Press, 1984), p. 37. Needler argues the same qualities exist in the medieval romance figure of the monstrous herdsman, but it is important that in Sidney's work it is man and not monster who is the super-beast: Needler, 'The Animal Fable', 426.
70. Sidney, *Defence of Poetry*, p. 53.
71. Jonathan D. Trigg, *Baptism in the Theology of Martin Luther* (Leiden: E. J. Brill, 1994), p. 106.
72. Stubbes, *Anatomie of Abuses*, sig.Biiii^r.
73. Sidney, *Old Arcadia*, p. 5.
74. Sidney, *Defence of Poetry*, p. 25.
75. Ben Jonson, *Volpone* (1606), in Herford and Simpson, ed., *Ben Jonson*, Volume V, 'The Epistle', lines 104–9.
76. Jonson, *Volpone*, 'Prologue', lines 4 and 8.
77. Ben Jonson, *Every Man In His Humour* (1598), in Herford and Simpson, ed., *Ben Jonson*, Volume III, I. v. 88–95.
78. Sidney, *Old Arcadia*, p. 315.
79. See Andrew Gurr, *Playgoing in Shakespeare's London* (Cambridge: Cambridge University Press, 1987), pp. 85–109.
80. Jonson, *Timber, or Discoveries*, in Herford and Simpson, ed., *Ben Jonson*, Volume VIII, line 2031.
81. See, for example, R. B. Parker, 'Volpone and Reynard the Fox', *Renaissance Drama*, n.s. 7 (1976), 3–42; D. A. Scheve, 'Jonson's *Volpone* and Traditional Fox Lore', *Review of English Studies*, n.s. 1 (1950),

242–4; and Malcolm H. South, 'Animal Imagery in *Volpone*', *Tennessee Studies in Literature*, 10 (1965), 141–50.

82. Sidney, *Old Arcadia*, p. 223.
83. Jonson, *Volpone*, I. ii. 82 and 85. The reference to the furs is repeated again at I. ii. 97 as if to reinforce the importance of the costume.
84. An important interpretation of the play which also looks at the theatricality of *Volpone* and which has informed my reading is Stephen J. Greenblatt, 'The False Ending in *Volpone*', *Journal of English and Germanic Philology*, 75 (1976), 90–104.
85. Jonson, *Volpone*, I. ii. 87–90.
86. Ibid., V. xii. 149–50.
87. Another example of the importance of the ideas of humanist educationists can be found in Jonson's *Epicoene*. Amorous La Foole's status as a poor and therefore animal interpreter is represented in his misremembering of *The Morall Philosophie of Doni*. La Foole thinks that it contains the story of Reynard, it does not, and his forgetfulness serves to underline the need for reiteration in learning. He may be casting his eyes over the right texts, but he is not reading them properly. Ben Jonson, *Epicoene* (1609), in Herford and Simpson, ed., *Ben Jonson*, Volume III, IV. iv. 76–8.
88. Jonson, *Volpone*, 'Epilogue', 2–4.
89. On this issue see Felix Raab, *The English Face of Machiavelli: A Changing Interpretation 1500–1700* (London: Routledge, 1964).
90. Niccolò Machiavelli, Letter 137, in *The Letters of Machiavelli*, translated by Allan Gilbert (New York: Capricorn, 1961), p. 142.
91. There are no copies of texts by Machiavelli in Jonson's library, as listed by David McPherson in 'Ben Jonson's Library and Marginalia: An Annotated Catalogue', *Studies in Philology*, 71: 5 (1974), but Jonson had clearly read Machiavelli's work. For a discussion of Jonson's relationship with Machiavelli, see Daniel C. Boughner, *The Devil's Disciple: Ben Jonson's Debt to Machiavelli* (New York: Philosophical Library, 1968).
92. In his study of Jonson and Lucian Douglas Duncan utilises this same Machiavellian moment in a very different way; see Duncan, *Ben Jonson and the Lucianic Tradition* (Cambridge: Cambridge University Press, 1979), p. 120.
93. Machiavelli, *The Letters*, p. 140.
94. T. H. White, ed., *The Book of Beasts: Being a Translation from a Latin Bestiary of the Twelfth Century* (1954), (Reprinted, Stroud: Alan Sutton, 1992), p. 54.
95. Duncan, *Ben Jonson and the Lucianic Tradition*, p. 128.
96. Ben Jonson, *The Alchemist* (1610), in Herford and Simpson, ed., *Ben Jonson*, Volume V, 'Prologve', line 3.
97. *Second Tome of Homilies*, fol.153ʳ; Jonson, *Volpone*, I. ii. 88–9.
98. Aulus Gellius, cited in Tony Davies, *Humanism* (London: Routledge, 1997), p. 126.
99. For an analysis of this animal image see Ian Donaldson, 'Jonson's Tortoise', in Jonas A. Barish, ed., *Volpone: A Casebook* (London: Macmillan, 1972), pp. 189–94.

100. Laura Levine, 'Men in Women's Clothing: Anti-theatricality and Effeminization from 1579 to 1642', *Criticism*, 28: 2 (1986), 126; reprinted in Levine, *Men in Women's Clothing: Anti-Theatricality And Effeminization, 1579–1642* (Cambridge: Cambridge University Press, 1994), pp. 10–25.

101. William Prynne, *Histrio-Mastix: The Players Scourge, or Actors Tragedie* (1633), p. 892.

102. Ibid.

103. Ibid.

104. Ibid., p. 893.

105. Calvin, *Institutes*, Volume II, p. 243.

106. [Munday], *Second and Third Blast*, p. 48.

107. Stubbes, *Anatomie of Abuses*, sig.Cvi^v.

108. Writing in the 1650s, the Quaker George Fox records that he was known as the 'man in leathern breeches', a title which refers, according to the editor of his *Journal*, to the fact that Fox wore 'a suit of leather, doublet and breeches.' In this Fox represents a return to 'pure' clothing (Stubbes' version), but the fact that Fox was known by his leather clothing also reveals a possibility of Prynne's line existing: Fox (appropriately named) is made animal by wearing animal skins. Fox, *The Journal of George Fox*, John L. Nickalls, ed. (London: Religious Society of Friends, 1975), pp. 83–4.

109. William Lamont records the appallingly appropriate story of the supporter of Prynne, Burton and Bastwick, who enacted her support for her maimed heroes soon after their public ear-cropping. She 'named her cats, Prynne, Burton and Bastwick, and then cut off their ears.' Lamont, *Marginal Prynne 1600–1669* (London: Routledge, 1963), p. 40.

110. Martindale, *English Humanism*, p. 19.

Chapter 4

1. Francis Bacon, *The Advancement of Learning* (1605), in James Spedding, Robert Leslie Ellis and Douglas Denon Heath, ed., *The Works of Francis Bacon* (1859; reprinted Stuttgart: Friedrich Frommann, 1963), Volume III, p. 346.

2. Bacon, 'The Plan' of *The Great Instauration* (1620), in Spedding *et al.*, ed., *Works*, Volume IV, p. 30.

3. Bacon, *Novum Organum* (1620), in Spedding *et al.*, ed., *Works*, Volume IV, p. 55.

4. Bacon, *The Great Instauration*, Volume IV, p. 30.

5. W. Rawley, 'To The Reader', preface to Bacon, *The New Atlantis* (1627), in Spedding *et al.*, ed., *Works*, Volume III, p. 127.

6. T. H. White, ed., *The Book of Beasts: Being a Translation from a Latin Bestiary of the Twelfth Century* (1954; reprinted Stroud: Allan Sutton, 1992), pp. 45–6.

7. Thomas Wright, 'Preface', in Wright, ed., *Alexandri Neckam: De Naturis Rerum* (London: Longman, 1863), p. lii; Michael R. Best and Frank H.

Brightman, ed., *The Book of Secrets of Albertus Magnus* (1550 edition), (Oxford: Clarendon, 1973), pp. 86 and 94.

8. Stephan Batman, *Batman vppon Bartholome, his Booke De Proprietatibus Rerum* (1582), fol.333ʳ. 'Iumenta' means beast of burden.

9. This categorising of animals in terms of their use to humans continued into the eighteenth century. Harriet Ritvo records the response to the kangaroo in 1770: '[Joseph] Banks and [Captain James] Cook concurred in proclaiming the otherwise unclassifiable new discovery "excellent food."' Ritvo, *The Platypus and the Mermaid and Other Figments of the Classifying Imagination* (Cambridge Mass.: Harvard University Press, 1997), p. 1.

10. John Henry has characterised a popular understanding of empiricism (which he does not share) as 'ideologically neutral, unbiased and objective'. Henry, 'The Scientific Revolution in England', in Roy Porter and Mikuláš Teich ed., *The Scientific Revolution in National Context* (Cambridge: Cambridge University Press, 1992), p. 199.

11. Charles E. Raven, *English Naturalists From Neckam to Ray: A Study in the Making of the Modern World* (Cambridge: Cambridge University Press, 1947), pp. 2–3.

12. For another interpretation of Bacon's language and modern historiography see Ronald Levao, 'Francis Bacon and the Mobility of Science', *Representations*, 40 (1992), 2.

13. Francis Klingender, *Animals in Art and Thought at the End of the Middle Ages* (London: Routledge, 1971), pp. 341 and 391–2. See also Brunsdon Yapp, 'Introduction' in Wilma George and Brunsdon Yapp, *The Naming of the Beasts: Natural history in the medieval bestiary* (London: Duckworth, 1991), p. 1.

14. On this issue see Joyce E. Salisbury, *The Beast Within: Animals in the Middle Ages* (London: Routledge, 1994), pp. 112–17. According to Allen Debus, the naturalists Conrad Gesner and Ulisse Aldrovandi, both writing in the mid-sixteenth century, pushed aside the bestiaries and began a scientific study of animals based on observation and experience. Debus, *Man and Nature in the Renaissance* (Cambridge: Cambridge University Press, 1978), p. 52.

15. Best and Brightman, 'Introduction' in Best and Brightman ed., *The Book of Secrets*, p. xii.

16. Ibid., p. xxx.

17. Edward Topsell, *The Historie of Fovre-Footed Beastes* (1607), pp. 711–21.

18. Topsell, 'The Epistle Dedicatory', *Historie of Fovre-Footed Beastes*, sig.A4ʳ.

19. Conrad Gesner, 'The First Epistle of Doct. Conradvs Gesnervs', in Topsell, *Historie of Fovre-Footed Beastes*, sig.¶1ᵛ.

20. Ibid., sig.¶2ᵛ.

21. For the importance of humanism on the development of natural philosophy, see Antonia McLean, *Humanism and the Rise of Science in Tudor England* (London: Heinemann, 1972).

22. Conrad Gesner, 'First Epistle', sig.¶1ᵛ.

23. Best and Brightman, ed., *Book of Secrets*, p. 3.

24. *The Secrets of Albertvs Magnvs. Of the Vertues of Hearbes, Stones, and certaine Beasts* (1617), sig.A2ᵛ. The changes in the prefatory material to *The Book of Secrets* are also pointed out in Best and Brightman, 'Introduction', p. xix.
25. Gesner, 'First Epistle', sig.¶2ʳ.
26. Richard D. French, 'Animal Experimentation: Historical Aspects', in Warren T. Reich ed., *Encyclopedia of Bioethics* (New York: Free Press, 1978), p. 76. For a useful historical overview of the development of vivisection, see Andreas Holger Maehle and Ulrich Tröhler, 'Animal Experimentation from Antiquity to the end of the Eighteenth Century: Attitudes and Arguments', in N. A. Rupke, ed., *Vivisection in Historical Perspective* (London: Routledge, 1990), pp. 14–47.
27. Carolyn Merchant, *The Death of Nature: Women, Ecology and the Scientific Revolution* (1980; reprinted London: Harper Collins, 1990), p. 169.
28. Ibid., p. 164.
29. Bacon, 'The Plan' of *The Great Instauration*, Volume IV, p. 30. The idea of the formative role of the nurse is touched on in a different way in Edmund Spenser's *A viewe of the presente state of Irelande* where he warns that children 'sucke even the nature and disposicion of their nurses'. In *Valentine and Orson* Orson is fed by the she-bear, and 'This Child, by reason of the nutriment it received from the Beare, became rough all over like a beast'. Spenser, *A viewe of the presente state of Irelande* (1596), in Rudolf Gottfried ed., *Spenser's Prose Works* (Baltimore: Johns Hopkins University Press, 1949), p. 119; *Valentine and Orson. The Two Sonnes of the Emperour of Greece* (1637), p. 26.
30. Bacon, *Novum Organum*, Volume IV, p. 93.
31. Bacon, *Thoughts and Conclusions on the Interpretation of Nature or a Science of Productive Works* (1604), in Benjamin Farrington, ed., *The Philosophy of Francis Bacon: An Essay on its Development from 1603 to 1609 with New Translations of Fundamental Texts* (Liverpool: Liverpool University Press, 1964), p. 74. This emphasis on the role of childhood in the formation of the human mind is repeated in many of Bacon's works: see, for instance, *Valerius Terminus of the Interpretation of Nature* (1603), in Spedding *et al.*, ed., *Works*, Volume III, p. 231.
32. See, for example, John Dod and Robert Cleaver, *A Treatise or Exposition Vpon the Ten Commandments* (1603), sig.7ʳ.
33. Leah Sinanoglou Marcus has noted that 'many intellectuals viewed the new science with alarm: to follow it they were obliged to undergo the painful process of cutting off their own mental roots, of wrenching apart a continuum from childhood belief to its adult elaboration.' Marcus, *Childhood and Cultural Despair: A Theme and Variations in Seventeenth-Century Literature* (Pittsburg: University of Pittsburg Press, 1978), pp. 92–3. For a discussion of the problematic place of the child and childhood in Bacon's ideas see my 'Calling Creatures by their True Names: Francis Bacon, The New Science and the Beast in Man', in Erica Fudge, Ruth Gilbert and Susan Wiseman, ed., *At the Borders of the Human: Beasts, Bodies and Natural Philosophy in the Early Modern Period* (Basingstoke: Macmillan, 1999).

34. Bacon, *Novum Organum*, Volume IV, p. 55.
35. Bacon, *Thoughts on Human Knowledge* (1604), in Farrington, ed., *Philosophy of Francis Bacon*, p. 41.
36. The emphasis on materiality, as well as the notion of progress is surely at the heart of Christopher Hill's alignment of Bacon and Marx. See Hill, *The Intellectual Origins of the English Revolution* (1965), (reprinted London: Granada, 1972), pp. 89–90.
37. Bacon, *The Masculine Birth of Time* (1603), in Farrington, ed., *Philosophy of Francis Bacon*, p. 69.
38. Bacon, *Novum Organum*, Volume IV, pp. 82 and 60.
39. Bacon, *Of the Wisdom of the Ancients* (1609), in Spedding *et al.*, ed., *Works*, Volume VI, p. 696.
40. Bacon, *Advancement of Learning*, Volume III, p. 345.
41. Paolo Rossi, *Francis Bacon: From Magic to Science*, translated by Sacha Rabinovitch (London: Routledge, 1968), pp. 127–8.
42. Bacon, *Preparative Towards a Natural and Experimental History* (1620), in Spedding *et al.*, ed., *Works*, Volume IV, p. 255.
43. Timothy H. Paterson, 'Bacon's Myth of Orpheus: Power as a Goal of Science in *Of the Wisdom of the Ancients*', *Interpretation*, 16: 3 (1989), 429.
44. Ibid., 430. His comments here are reminiscent of Roger Chartier's ideas cited in Chapter 3.
45. Bacon, *Wisdom of the Ancients*, Volume VI, p. 695.
46. Paterson, 'Bacon's Myth of Orpheus', 430.
47. Bacon, *Wisdom of the Ancients*, Volume VI, p. 698.
48. Maurice Slawinski has shown how the new scientific endeavour also used some of the ideas of rhetoric – the 'unscientific' – in its creation, something which 'remind[s] us that the new philosophy itself was built on metaphysical foundations outside the bounds of proof'. Slawinski, 'Rhetoric and science/rhetoric of science/science of rhetoric', in Stephen Pumfrey, Paolo L. Rossi and Maurice Slawinski, ed., *Science, culture and popular belief in Renaissance Europe* (Manchester: Manchester University Press, 1991), p. 88.
49. Bacon, *Masculine Birth of Time*, p. 62.
50. David Hawkes has argued that in *Novum Organum* can be traced 'a fully secular theory of false consciousness.' I want to argue that there is nothing secular at all in Bacon's natural philosophy. Hawkes, *Ideology* (London: Routledge, 1996), p. 31.
51. Bacon, *Valerius Terminus*, Volume III, p. 217.
52. A link might be made here to Bacon's understanding of the centrality of the monarch in the law which was one source of difference between Bacon and Coke. See Richard Helgerson, *Forms of Nationhood: The Elizabethan Writing of England* (Chicago: Chicago University Press, 1992), pp. 73–4.
53. On the issue of vegetarianism John Calvin cites a number of contradictory moments from scripture which both support and undermine the suggestion and then concludes 'Therefore I thinke it shall be better if we say nothing concerning the same.' Calvin, *A Commentarie*

of John Calvin, vpon the first booke of Moses called Genesis, translated by Thomas Tymme (1578), p. 48.

54. Henry Holland, *The Historie of Adam, or the foure fold state of Man* (1606), sig.Biii^r.
55. Bacon, *Valerius Terminus*, Volume III, p. 222.
56. Henry Arthington, *Principall Points of holy profession* (1607), sig.B2^v.
57. [Thomas Morton], *A Treatise of the threefolde state of man* (1596), p. 24.
58. Calvin, *Commentarie*, p. 75. Matthew Senior, calls this a moment of 'linguistic perfection'. Senior, '"When the Beasts Spoke": Animal Speech and Classical Reason in Descartes and La Fontaine', in Jennifer Ham and Matthew Senior, ed., *Animal Acts: Configuring the Human in Western Culture* (London: Routledge, 1997), p. 66.
59. Arthington, *Principall Points*, sig.B2^r.
60. Alister E. McGrath has noted how 'Calvin gave a new religious motivation to the scientific investigation of nature'. Charles Webster has noted that the anti-authoritarian (anti-Aristotelean) nature of the new science sat comfortably with the Reformed belief in 'personal revelation'. McGrath, *Reformation Thought: An Introduction*, Second Edition (Oxford: Blackwell, 1993), p. 231; Webster, *The Great Instauration: Science, Medicine and Reform 1626–1660* (London: Duckworth, 1975), p. 189; see also p. 506.
61. Bacon, *Valerius Terminus*, Volume III, p. 221.
62. Thomas Adams, *Meditations Vpon Some Part of the Creed*, in *The Workes of Tho: Adams* (1630), p. 1132.
63. The term is Margarita Bowen's. She also notes that the term 'empiricism' come from the Greek term *empeiria*, which means experience. Bowen, *Empiricism and Geographical Thought: From Francis Bacon to Alexander von Humboldt* (Cambridge: Cambridge University Press, 1981), pp. 4 and 3.
64. On the issue of the classifying of flora, and the renaming of plants, see Keith Thomas, *Man and the Natural World: Changing Attitudes in England 1500–1800* (London: Penguin, 1984), pp. 51–91.
65. Bacon, *Novum Organum*, Volume IV, p. 47.
66. In his poem 'To the Royal Society' Abraham Cowley avoids the Bacon/Christ problem by referring to Bacon as 'Moses'. Cowley, 'To the *Royal Society*' in Thomas Sprat, *The History of the Royal Society* (1667), sig.B^r.
67. In *defence* of the biblical representation of humanity's relation with the natural world, F. B. Welbourn states, 'Man is to rule the rest of creation; and the experience of being ruled may not always be pleasant.' Welbourn, 'Man's Dominion', *Theology*, 78: 665 (1975), 563. This is written in response to Lynn White Jr's seminal article 'The Historical Roots of Our Ecological Crisis', *Science*, 155: 3767 (1967), 1203–7. The General Synod of the Church of England proposes that 'this Synod, affirming its belief and trust in God the Father who made the world, believe that the dominion given to human beings over the natural order is that of stewards who have to render an account'. The General Synod of the Church of England, 'What is the Church's

View?' (Board for Social Responsibility, 1994). I am grateful to David Skidmore for sending me a copy of this booklet.

68. Jonathan Sawday, *The Body Emblazoned: Dissection and the Human Body in Renaissance Culture* (London: Routledge, 1995), p. 106.

69. Joseph Fletcher, *The Historie of the Perfect-Cursed Blessed Man* (1629), p. 7.

70. I am not arguing that vivisection is no longer a massively problematic part of culture, but I am looking here specifically at the problems faced by seventeenth-century thinkers.

71. Gesner, 'First Epistle', sig. ¶5ʳ.

72. Bacon, *New Atlantis*, Volume III, p. 159. Bacon died as a result of a chill caught after freezing a chicken to see if 'flesh might not be preserved in snow, as in salt.' John Aubrey, *Brief Lives*, cited in Anthony Quinton, *Francis Bacon* (Oxford: Oxford University Press, 1980), pp. 6–7.

73. Bacon, 'The Plan' of *The Great Instauration*, Volume IV, p. 30.

74. William Harvey, *An Anatomical Disputation Concerning the Movement of the Heart and Blood in Living Creatures* (1628), Gweneth Whitteridge, ed. (Oxford: Blackwell, 1976), p. 7.

75. See F. J. Cole, 'Harvey's Animals', *Journal of the History of Medicine*, 12 (1957), pp. 106–7.

76. Harvey, *Anatomical Disputation*, p. 32.

77. Bacon, *Masculine Birth of Time*, p. 69; and Bacon, *Novum Organum*, Volume IV, p. 93. For a discussion of the changing place of the scientist in the early modern period see Jonathan Sawday, 'The Fate of Marsyas: Dissecting the Renaissance Body', in Lucy Gent and Nigel Llewellyn ed., *Renaissance Bodies* (London: Reaktion, 1990), pp. 111–35; and Mario Biagioli, 'Scientific Revolution, social bricolage, and etiquette', in Porter and Teich, ed., *Scientific Revolution in National Context*, especially p. 18.

78. William Prynne was writing five years after the publication of Harvey's discovery. For a discussion of the impact of science on Prynne's ideas see my 'Temples of God: William Prynne and the New Science', in Tracey Hill and Jeffrey Rodman, ed., *The Body of Truth: Corporeality and Power in Early Modern Culture* (Bath: Sulis Press, 1999).

79. Bacon, *Masculine Birth of Time*, p. 72.

80. Marion Trousdale, *Shakespeare and the Rhetoricians* (London: Scolar Press, 1982), p. 25.

81. Bacon, *Valerius Terminus*, Volume III, p. 222.

82. Carolyn Merchant underlines the importance of this exclusion – and literal seclusion – of women in *The New Atlantis*; Merchant, *Death of Nature*, p. 174.

83. Bacon, *New Atlantis*, Volume III, p. 135.

84. Ibid., pp. 147–51.

85. Julian Martin, *Francis Bacon, The State and the Reform of Natural Philosophy* (Cambridge: Cambridge University Press, 1992), p. 165.

86. Bacon, *Novum Organum*, Volume IV, p. 114.

87. The linguistic disorder of Babel is cancelled with the arrival of Christianity: 'every one read upon the Book and Letter, as if they had

been written in his own language.' Bacon, *New Atlantis*, Volume III, pp. 138–9.

88. William Rawley, 'To The Reader', preface to Bacon, *New Atlantis*, Volume III, p. 127.
89. Ibid., p. 129.
90. Thomas Scot pre-empted Bacon when he called travellers 'lyers'. Tho: Scot, *Phylomythie or Philomythologie wherin Outlandish Birds, Beasts, and Fishes, are Taught to Speak true English plainely* (1616), sig.¶ᵛ.
91. See Stephen Greenblatt, *Marvellous Possessions: The Wonder of the New World* (Oxford: Clarendon, 1991), p. 33.
92. *The Fables of Esope in English* (c.1570), title page.
93. Salisbury, *Beast Within*, p. 106.
94. Louis Marin, 'The "Aesop" Fable-Animal', in Marshall Blonsky, ed., *On Signs* (Oxford: Blackwell, 1985), p. 335.
95. Ibid., p. 337. In *Valentine and Orson* there is a similar resort to physical objects to make meaning: Orson 'threw the pot against the ground, making a signe vnto *Valentine* to draw him more.' *Valentine and Orson*, p. 63.
96. *Fables of Esope in English*, sig.Aiiʳ.
97. Ibid., sig.Aiiiʳ⁻ᵛ.
98. Ibid., sig.Aiiiᵛ.
99. Annabel Patterson, *Fables of Power: Aesopian Writing and Political History* (London: Duke University Press, 1991), p. 74.
100. Marin, '"Aesop" Fable-Animal', p. 336.
101. Carla Mazzio has called this, in a wonderfully embodied image, the 'slipperiness' of the tongue. Mazzio, 'Sins of the Tongue', in David Hillman and Carla Mazzio, ed., *The Body in Parts: Fantasies of Corporeality in Early Modern England* (London: Routledge, 1997), p. 54.
102. *Fables of Esope in English*, sig.Bvʳ⁻ᵛ.
103. Ibid., sig.Bviʳ.
104. On the ambivalence of the tongue see Mazzio, 'Sins of the Tongue'.
105. Bacon, *New Atlantis*, Volume III, p. 137.
106. Bacon, *Thoughts on Human Knowledge*, p. 41.
107. For a discussion of the importance and religious place of 'the matter of fact' in English scientific thought, see Henry, 'Scientific Revolution in England', pp. 196–8.
108. Ibid., p. 199.
109. To return to the argument made in the Introduction, Emmanuel Levinas, writing of his and his fellow Jews' experiences in the Labour Camp, says that the Nazi Guard's eyes 'stripped us of our human skin'. Levinas, 'The Name of a Dog, or Natural Rights', in *Difficult Freedom: Essays on Judaism*, translated by Séan Hand (London: Athlone Press), p. 153.

Chapter 5

1. Pierre Le Loyer, *A Treatise of Specters or straunge Sights, Visions and Apparitions appearing sensibly vnto men* (1605), fol.107ʳ. La Loyer tells the story as evidence of what Ambroise Paré has called the 'efficacy

of imagination'. Paré, *Of Monsters and Prodigies* (1573), in *The Works of that Famous Chirurgeon Ambrose Parey* (1678), p. 585.

2. Le Loyer, *Treatise of Specters*, fol.107^{r-v}.
3. Ibid., fol.107^v.
4. Ibid., fol.109^{r-v}.
5. Ibid., fol.109^{r-v}.
6. Edward Coke, *The First Part of the Institutes of the Lawes of England* (1628), fol.7^v.
7. Ibid., fol.8^r.
8. Sir John Davies, *Le Primer des Cases ... en Ireland* (1615), cited in Richard Helgerson, *Forms of Nationhood: The Elizabethan Writing of England* (Chicago: University of Chicago Press, 1992), p. 87.
9. Alan Harding, *A Social History of the English Law* (Harmondsworth: Penguin, 1966), p. 30.
10. Edward Coke, 'First Part' of *Les Reports* (1600; reprinted 1609), sig. ⁊ iii^r.
11. See Edward Coke, *Quinta Pars Relationum Edwardi Coke* (1605; reprinted 1607), sig.Av^v–Avi^r.
12. Coke, cited in Helgerson, *Forms of Nationhood*, p. 99.
13. John Underwood Lewis, 'Sir Edward Coke (1552–1633): His Theory of "Artificial Reason" As a Context for Modern Basic Legal Theory', *Law Quarterly Review*, 84: 335 (1968), 334.
14. Rather than proposing that humanism had an impact on the systematisation of the law in the early seventeenth century, J. H. Baker also suggests, in a footnote, a reversal of this model: '[i]t may be that "humanism" itself should be traced to the new spirit of enquiry among lawyers'. Baker, 'English Law and the Renaissance', in Baker, *The Legal Profession and the Common Law* (London: Hambledon, 1986), p. 476.
15. Paul H. Kocher, 'Francis Bacon on the Science of Jurisprudence', *Journal of the History of Ideas*, 18: 1 (1957), 3.
16. Whether achieving benefit of clergy actually represented literacy is, as Cynthia B. Herrup notes, unlikely: Herrup, *The Common Peace: Participation and the Common Law in Seventeenth-Century England* (Cambridge: Cambridge University Press, 1987), p. 48.
17. Edward Coke, *La Sept Part Des Reports* (1608), sig.A5^v; Coke, First Part, *Les Reports*, sig. ⁊ iii^r.
18. Ibid., sig. ⁊ iii^r.
19. Edward Coke, *La Dix^{me} Part des Reports* (1614), sig.ciij^r.
20. Thomas North, 'The Prologue', in *The Morall Philosophy of Doni* (1570; reprinted 1601) sig.B1^v.
21. Richard Helgerson proposes that legal reports forbid 'access to an identifiable origin of author', but other claims question this idea of the anonymity of the law: J. H. Baker has proposed that the author is not dead within the law until the late nineteenth century; 'the most learned lawyer today would not know (or think it useful to know) who reported, say, *Donaghue v. Stephenson*.' Helgerson, *Forms of Nationhood*, p. 87; Baker, *An Introduction to English Legal History*, Third Edition (London: Butterworths, 1990), p. 211.

22. For a discussion of the sources of Coke's *Reports*, see J. H. Baker, 'Coke's Notebooks and the Sources of his Reports', in Baker, *The Legal Profession*, pp. 177–204. The *Reports* first appeared in French, the official language of the law, although the prefatory material is in both Latin and English. The main text was first published in English in 1658.

23. Coke, 'First Part', sig. ⁊ᛆiv^r.

24. For a record of these trials see E. P. Evans, *The Criminal Prosecution and Capital Punishment of Animals: The Lost History of Europe's Animal Trials* (1906; reprinted London: Faber and Faber, 1988).

25. Esther Cohen, 'Law, Folklore and Animal Lore', *Past and Present*, 110 (1986), 14.

26. Peter Mason, 'The excommunication of caterpillars: ethno-anthropological remarks on the trial and punishment of animals', *Social Science Information*, 27: 2 (1988), 267.

27. J. A. Sharpe, *Crime in Early Modern England 1550–1750* (London: Longman, 1984), p. 12.

28. J. J. Finkelstein, 'The Ox That Gored', *Transactions of the American Philosophical Society*, 7: 2 (1981), 28.

29. Cohen, 'Law, Folk Lore and Animal Lore', 36.

30. E. P. Thompson, 'Patrician Society, Plebian Culture', *Journal of Social History*, 7 (1974), 382–405.

31. Pieter Spierenberg, *The Spectacle of Suffering: Executions and the Evolution of Repression: From a Preindustrial Metropolis to the European Experience* (Cambridge: Cambridge University Press, 1984), p. 201.

32. See W. W. Hyde, 'The Prosecution and Punishment of Animals and Lifeless Things in the Middle Ages and Modern Times', *University of Pennsylvania State Law Review*, 64 (1916), 696–8. Hyde's inclusion of the trial of stones with his discussion of the trial of animals is further evidence of the fact that the trials do not imply animal sentience.

33. See Evans, *Criminal Prosecution*, p. 140.

34. Nicholas Humphrey, 'Foreword' to Evans, *Criminal Prosecution*, p. xxii.

35. Finkelstein, 'The Ox that Gored', 72.

36. Edward Coke, *The Third Part of the Institutes of the Laws of England* (1644), p. 57.

37. Michael Dalton, *The Covntrey Jvstice* (1618), p. 218.

38. Coke, *Third Part of the Institutes*, p. 57.

39. Case cited in F. G. Emmison, *Elizabethan Life: Disorder* (Chelmsford: Essex County Council, 1970), p. 227.

40. Sir George Croke, 'The Lord Chandois Case' (1619), in *The Second Part of the Reports of Sir George Croke* (1683), p. 483.

41. The use of deodand came to an end in 1846: Baker, *Introduction to English Legal History*, p. 437.

42. W. S. Holdsworth, *A History of the English Law* (London: Methuen, 1924), Volume II, p. 47.

43. S. F. C. Milsom, *Historical Foundations of the Common Law*, Second Edition (London: Butterworths, 1981), p. 311.

44. Sir George Croke, 'Boulton *versus* Banks' (1632), in *The Reports of Sir George Croke* (1658), p. 254.
45. Dalton, *Covntrey Jvstice*, p. 234.
46. William Lambard, *Eirenarcha: or of the Offices of Justices of the Peace in foure Bookes* (1592), p. 268. Dalton repeats Lambard's designation in *Covntry Jvstice*, p. 233.
47. Dalton, *Covntrey Jvstice*, p. 234.
48. Francis Bacon, *Valerius Terminus of the Interpretation of Nature* (1603), in James Spedding, Robert Leslie Ellis and Douglas Denon Heath ed., *The Works of Francis Bacon* (1859; reprinted, Stuttgart: Friedrich Frommann, 1963), Volume III, p. 222.
49. Croke, 'Sir Francis Vincent *versus* Lesney' (1626), in *Reports*, pp. 18–19.
50. Croke, 'Sir Martyn Lister *versus* Horne' (1640), in *Reports*, pp. 544–5.
51. Coke, *Third Part of the Institutes*, p. 98.
52. Ibid.
53. Baker, *Introduction to English Legal History*, p. 428.
54. Leonard Mascall, *The first Booke of Cattell* (1591), sig.O1^v.
55. Mascall's text includes various 'remedies to helpe most diseases as may chaunce vnto' sheep, goats, pigs and dogs. In this sense it fulfils Conrad Gesner's belief that 'euery heard-man and leach in the fields' knows 'profitable medicines' which can be utilised in the (more important) study and cure of humanity. Mascall, *The first Booke of Cattell*, sig.O1^v; Gesner, 'The First Epistle of Doct. Conradvs Gesnervs', in Edward Topsell, *The Historie of Foure-Footed Beastes* (1607), sig.¶1^v.
56. Lambard, *Eirenarcha*, p. 267. The list appears almost verbatim in Dalton, *Covntrey Jvstice*, p. 233.
57. This is a practice known as 'agistment': 'taking another man's stock in to graze for a money payment or share in the progeny of that stock.' Peter J. Bowden, *The Wool Trade in Tudor and Stuart England* (London: Macmillan, 1962), p. 15.
58. East Sussex Record Office (hereafter referred to as ESRO), Q/R/E 58 (84, 85, 86, and 88), (1642). In this section of the chapter I am using depositions from the period 1642–9 primarily because of the thorough work done on depositions from 1594–5, 1614–18, 1625–28 and 1636–40 in the ESRO by Cynthia B. Herrup and discussed in *The Common Peace*. The similarity of the records dealing with the theft of animals across the whole period is significant and represents a stability in notions of ownership and in the relation with the animal.
59. See B. C. Redwood, ed., *Quarter Sessions Order Book, 1642–1649* (Lewes: Sussex Record Society, 1954).
60. The branding and ear-clipping of human criminals links them with the animal and is another form of degradation. Prynne had his ears clipped twice, Jonson was threatened with ear-clipping. S. R. Gardiner ed., *Documents Relating to the Proceedings Against William Prynne in 1634 and 1637* Camden Society, n.s. 18 (1877); Richard Dutton, *Ben Jonson*, cited in Peter Stallybrass and Allon White, *The Politics and Poetics of Transgression* (Ithaca: Cornell University Press, 1986), p. 74.

61. ESRO, Q/R/E 27 (101–3), cited in Herrup, *Common Peace*, p. 78.
62. ESRO, Q/R/E 70 (72), (1645).
63. ESRO, Q/R/E 67 (74), (1644).
64. ESRO, Q/R/E 68 (61), (1645).
65. I am not interested in the economic reasons for the theft of animals here, but it is worth noting that the loss to the owner of a sheep was a very real one. As Bowden has noted, the annual wool-clip of an average wether (gelded male) sheep was worth up to one-third of the value of the animal itself. Bowden, *Wool Trade*, p. 3. The distinction here between object and subject also reflects the need to make a distinction between the animal and the meat: the individual and the ways in which it was exploited, a point to which I return.
66. ESRO, Q/R/E 67 (71), (1644). This case offers an illustration of the way in which the owner of the animal is responsible for that animal's actions: both John Kent, the owner of the dog, and his brother, Richard Kent, who assisted him in covering up the crime, were found guilty: both were whipped. See Redwood, *Quarter Sessions Order Book*, p. 75. Herrup records a number of similar cases where the carcass and not the animal is recovered: ESRO, Q/R/E 12 (30); ESRO, Q/R/E 36 (50); ESRO, Q/R/E 46 (45 and 56); ESRO, Q/R/E 35 (31); in Herrup, *Common Peace*, pp. 75, 76, 77 and 84.
67. Only the conviction of 'Nicholas Leany alias Barden' is recorded in the extant indictments, Sage is not mentioned. Redwood, *Quarter Sessions Order Book*, p. 169.
68. ESRO, Q/R/E 82 (64), (1648).
69. ESRO, Q/R/E 64 (60), (1644). For the emphasis on cooking the animal as quickly as possible see also ESRO Q/R/E 46 (45 and 56), ESRO Q/R/E 35 (31) and Q/R/E 36 (103), in Herrup, *Common Peace*, pp. 77 and 84.
70. The distinction in English between the animal and its meat – sheep/mutton, cow/beef, pig/pork, calf/veal – is not present in other languages. In French, for instance, mouton means both mutton and sheep, boeuf: beef and cow, porc: pork and pig, veau: veal and calf. The English term for the animal is not altered or used to represent meat, instead a new term, based on the French, is added to the language and a separation is made between the animal and the meat; the subject and the object. What is also made is a separation between the classes: the introduction of French terms with the Norman Conquest meant that those who tended and worked with the animal would use one (English) name, while those who consumed the animal would use another (French) one.
71. Thomas Adams, *Meditations Vpon Some Part of the Creed*, in *The Workes of Tho: Adams* (1630), p. 1132. For a discussion of the earlier part of this sermon see Chapter 4 above.
72. John Moore, *A Mappe of Mans Mortalitie* (1617), p. 40.
73. In his important study of the meaning of meat Nick Fiddes argues that 'It is fitting that … meat should have risen in both quantity consumed and in significance from around the seventeenth century onwards, at a time when science was increasingly stressing the need

to dominate nature'. I am not sure that Reformed angst and an
increase in meat-eating are necessarily totally at odds with each
other. Fiddes, *Meat: A Natural Symbol* (London: Routledge, 1991),
pp. 226–7. See also Colin Spencer, *The Heretic's Feast: A History of
Vegetarianism* (London: Fourth Estate, 1993).

74. John Rawlinson, *Mercy to a Beast. A Sermon Preached at Saint Maries
Spittle in London on Tuesday in Easterweeke, 1612* (Oxford: 1612), p. 34.
For a history of the profession see Sandra Billington, 'Butchers and
Fishmongers: Their Historical Contributions to London's Festivity',
Folklore, 101: 1 (1990), 97–103.

75. For a later discussion, see Robert Boyle, *Some Considerations Touching
the Usefulness of Experimental Natural Philosophy* (1660–63), in *The Works
of the Honourable Robert Boyle in Six Volumes* (1772), Volume II, esp.
pp. 6–18.

76. The Boyle Papers, Royal Society Library, London, Volume XXXVII,
fols. 186–193, reprinted in Malcolm R. Oster, 'The "Beame of
Diuinity": Animal Suffering in the Early Thought of Robert Boyle',
British Journal for the History of Science, 22: 2 (1989), 173–9, quote, 177.

77. Roger Crab, *The English Hermite, Or Wonder of this Age* (1655), pp. 3–4.
There is legal sense in Crab's argument; the 'common law divided
criminal responsibility equally between a thief and his receivers; the
convicted accomplice could hang along with the principal.' Herrup,
Common Peace, p. 82.

78. See Keith Thomas, *Man and the Natural World: Changing Attitudes in
England 1500–1800* (London: Penguin, 1984), p. 95.

79. On this issue see Marc Shell, 'The Family Pet', *Representations*, 15
(1986), 121–53.

80. Lambard, *Eirenarcha*, p. 268. Four breeds of dog are recognised as
property by the early modern law, as they are considered to be
working animals: these are, mastiffs, hounds, spaniels and tumblers.
Other dogs by implication are excluded.

81. Dalton, *Covntrey Jvstice*, p. 235.

82. On nineteenth-century attitudes see Kathleen Kete, *The Beast in the
Boudoir: Petkeeping in Nineteenth-Century Paris* (Berkeley: University of
California Press, 1994).

83. Thomas, *Man and the Natural World*, p. 113.

84. John Taylor, *Wit and Mirth* (1630) in W. Carew Hazlitt, ed., *Shakespeare
Jest Books*, (London: Willis and Southeran, 1864), Volume III, p. 35.

85. Sir Edmund Verney records an over-abundance of apes in England in
1636 in a letter to his son: 'A merchant of Lundun wrote to a factor of
his beyoand sea, desired him by the next shipp to send him 2 or 3
Apes; he forgot the r, and then it was 203 Apes. His factor has sent
him fower scoare, and sayes hee shall have the rest by the next shipp
… if y^r self or frends will buy any to breede on, you could never have
had such a chance as now.' Cited in Robert D. Altick, *The Shows of
London* (London: Bellknapp Press, 1978), p. 37.

86. Topsell, *Historie of Foure-Footed Beastes*, pp. 171–2, from John Caius, *Of
Englishe Dogges: The diversities, the names, the natures, and the Properties*,
translated by Abraham Fleming (1576), pp. 20–1.

87. Topsell, *Historie of Foure-Footed Beastes*, p. 163.
88. Ibid., p. 171. As mentioned above in Chapter 1, Dorothy Leigh saw the lascivious woman as a beast.
89. Keith Thomas, *Religion and the Decline of Magic* (1971; reprinted London: Penguin, 1991), p. 530.
90. George Gifford, *A Dialogve concerning Witches and Witchcraftes* (1593), sig.C^r.
91. Case cited in James Serpell, *In the Company of Animals: A Study of Human-Animal Relationships* (Oxford: Blackwell, 1986), pp. 45–6.
92. Elaine V. Beilin notes of literary attacks on women in the early modern period that they would 'point out that Eve listened to Satan and thus initiated all of humankind's future woe. Since then women had followed their guilty fore-mother by being disobedient, talkative, lascivious shrews.' Beilin, *Redeeming Eve: Women Writers of the English Renaissance* (Princeton N.J.: Princeton University Press, 1987), p. xviii.
93. Gervase Markham, *Countrey Contentments* (1615), pp. 88–9. Markham's use of the term 'jumpe' is echoed in later depositions dealing with bestiality where the term is also used as a euphemism for the sexual act. See, Public Record Office (hereafter referred to as PRO) ASSI 45/1/4 (17), (1642); ASSI 45/9/3 (85), (1670); ASSI 45/12/2 (90), (1678); and ASSI 45/13/2 (1), (1682). There are few surviving assize depositions extant from seventeenth century, so in the discussion of bestiality I will use depositions from the North Eastern Circuit. There are seventeen depositions dealing with bestiality from the period 1642 to 1689, and the similarities between the depositions, like the similarities in the depositions dealing with sheep theft, argue for their inclusion in this chapter.
94. William Shakespeare, *The Taming of the Shrew* (1594), IV. i. 176–82, in Stanley Wells and Gary Taylor, ed., *William Shakespeare: The Complete Works* (Oxford: Clarendon, 1988).
95. The limitations on owning a hawk and human social status are in Coke, *Third Part of the Institutes*, p. 98.
96. George Wilson, *The Commendation of Cockes and Cock-fighting* (1607), sig.C^r. See also Gervase Markham, *The Whole Art of Husbandrie* (1631), pp. 313–15.
97. In the extant depositions from the Northern Circuit Assizes, and in cases recorded in the Home Circuit Assizes during the reigns of Elizabeth I and James I no female bestialist is recorded. Notions of female bestialists, it is interesting to note, appear more often than not in fictional cases: in the male imagination. See, for example, Edward Fenton, *Certain Secrete wonders of Nature* (1569), p. 130; and William Turner, *A Compleat History of the Most Remarkable Providences* (1697), p. 26.
98. See John Canup, '"The Cry of Sodom Enquired Into": Bestiality and the Wilderness of Human Nature in Seventeenth-Century New England', *Proceedings of the American Antiquarian Society*, 98 (1988), 123.
99. Paré, *Of Monsters and Prodigies*, pp. 585 and 599. In his defence, the monstrous son, whose legal case began this chapter, makes a clear distinction between monsters born because of the actions of the

operation of the mother's imagination, and monsters 'which are borne of some beast'. These latter monsters, he admits, 'may lawfully be slaine'. Le Loyer, *Treatise of Specters*, fol.109^{r-v}.

100. Joyce E. Salisbury, *The Beast Within: Animals in the Middle Ages* (London: Routledge, 1994), pp. 84–101; Danielle Jacquart and Claude Thomasset, *Sexuality and Medicine in the Middle Ages* (Cambridge: Polity Press, 1988), p. 163.

101. Turner, *Compleat History*, p. 25.

102. PRO ASSI 45/5/3 (26), (1656).

103. Dalton, *Covntrey Jvstice*, p. 242.

104. Fenton, *Certaine Secrete wonders of Nature*, p. 130. Sodomy and buggery are often interchangeable terms in this period.

105. Evans, *Criminal Prosecution*, p. 147.

106. When Edmund Spenser writes of the Irish native's 'loathly filthines which is not to be named' he is, surely, hinting at bestial sexual practices. Spenser, *A viewe of the presente state of Irelande* (1596), in Rudolf Gottfried, ed., *Spenser's Prose Works* (Baltimore: Johns Hopkins, 1949), p. 102.

107. Coke, *Third Part of the Institutes*, p. 58.

108. Herrup, *Common Peace*, p. 3.

109. Ibid., p. 191.

110. John Calvin, *The Institutes of the Christian Religion* (1559 edition), translated by Henry Beveridge (London: James Clarke & Co.: 1949), Volume I, p. 243.

111. William Perkins, *A Discourse of Conscience* (1596), in *The Workes Of That Famovs and Worthy Minister of Christ in the Vniuersitie of Cambridge, Mr William Perkins* (1616–18), Volume I (1616), p. 523. This division into willed and unwilled sin might offer some explanation of the status of the atheist in Perkins' work (discussed in Chapter 2). There is a willed and an unwilled removal from God. The possibility of wilful removal still, however, raises the issue of Pelagianism.

112. Bracton, cited in Nigel Walker, *Crime and Insanity in England: Volume One: The Historical Perspective* (Edinburgh: Edinburgh University Press, 1968), p. 29.

113. Edward Coke, *The Second Part of the Institutes*, cited in Christopher Hill, *The Intellectual Origins of the English Revolution* (1965; reprinted, London: Granada, 1972), p. 248.

114. Cited in Dudley Wilson, *Signs and Portents: Monstrous Births from the Middle Ages to the Enlightenment* (London: Routledge, 1993), pp. 56–7.

115. PRO ASSI 45/12/2 (90), (1678).

116. A third interpretation of the problem of the primacy of intention and action, slightly different to the legal notions, comes from the Reformed sense of the inevitability of sin: Anthony Munday writes, 'For when necessitie onlie makes an euil thing to be left vndone, the verie desire of a filthie thing is condemned in such sort as if it were done.' [Munday], *A Second and Third Blast of Retrait from Plaies and Theaters* (1580), pp. 24–5.

117. Laurent Bonchel (1559–1629), *Biblioteque du Droit Francais*, cited in Gaston Dubois-Desaulle, *Bestiality: An Historical, Medical, Legal and*

Literary Study (c.1905), translated by A. F. N. (New York: Panurge Press, 1933), pp. 93–4.

118. William Perkins, *A Treatise Tending unto a Declaration Whether a Man be in the Estate of Damnation or in the Estate of Grace* (1596), in *Workes*, Volume I, p. 384. This is discussed in more detail in Chapter 2.

119. Edward Coke, 'Beverleys Case' (1625), in *The Reports of Sir Edward Coke* (1658), Part IV, p. 334.

120. See Anthony Michael Platt and Bernard L. Diamond, 'The Origins and Development of the "Wild Beast" Concept of Mental Illness and Its Relation to Theories of Criminal Responsibility', *Journal of the History of the Behavioral Sciences*, 1: 4 (1965), 355–67.

121. Coke, 'Beverleys Case', p. 335.

122. Dalton, *Covntrey Jvstice*, p. 215.

123. Coke, *Third Part of the Institutes*, p. 4. The translation of the Latin is from Walker, *Crime and Insanity*, p. 197.

124. There were also financial reasons why capital punishment was not considered to be appropriate for idiots. The family of the executed criminal lost their inheritance. Coke, 'Beverleys Case', p. 335.

125. Ibid.

126. Ibid.

127. Ibid. and 336–7.

128. Baker, *Introduction to English Legal History*, p. 275.

129. Coke, *Third Part of the Institutes*, p. 62.

Chapter 6

1. Throughout this chapter I am concentrating specifically on the work of Richard Overton. Texts by other Levellers – John Lilburne, William Walwyn and John Wildman – are not discussed because Overton's ideas are distinctive in their response to some of the debates about animals with which this book is concerned.

2. Throughout this discussion I am using the first, uncorrected edition of *Mans Mortallitie* (dated by Thomason as 19 January 1643/4).

3. The title page of the first edition dates it as 1643, but Thomason's exact dating would mean that the year was 1644 under the new system.

4. P. Zagorin, 'The Authorship of *Mans Mortallitie*', *The Library*, Fifth Series, 5: 3 (1950), 179–83, quotation, 181.

5. The reference is in [Richard Overton], *The Arraignment of Mr Persecution* (1645), in William Haller ed., *Tracts on Liberty in the Puritan Revolution 1638–1647* (New York: Columbia University Press, 1934), p. 230. Nigel Smith sees Overton's Mar-Priest tracts as 'partly motivated by the hostile response accorded to his defence of the mortalist heresy, *Mans Mortallitie*'. Smith, *Literature and Revolution in England 1640–1660* (London: Yale University Press, 1994), p. 302. For the repetition of ideas compare *Mans Mortallitie*, p. 17 with Overton, *An Appeale From the degenerate Representative Body* (1647), in Don. M. Wolfe, ed., *Leveller Manifestoes of the Puritan Revolution* (1944;

reprinted, London: Frank Cass, 1967), p. 158. I discuss this in more detail below. Zagorin also notes the significance of Richard Overton's access to a printing press in attributing *Mans Mortallitie*. Zagorin, 'Authorship', 181.

6. I am not concerned with the enlarged version of the text here, but Don M. Wolfe has suggested that it was written by Overton in collaboration with Milton. See Wolfe, 'Lilburne's Note on Milton', *Modern Language Notes*, 56: 5 (1941), 360.

7. Richard Overton, *Mans Mortallitie* (1643), p. 4.

8. Ibid., p. 7.

9. Ibid., p. 9.

10. Ibid., p. 21.

11. Ibid., p. 22. This point in itself was enough to place *Mans Mortallitie* beside 'John Milton's tract on divorce and Roger Williams' *Bloody Tenet of Persecution* as among the most scandalous yet seen in England.' Murray Tolmie, *The Triumph of the Saints: The Separate Churches of London 1616–1649* (Cambridge: Cambridge University Press, 1977), p. 82. See also H. N. Brailsford, *The Levellers and the English Revolution*, Christopher Hill, ed. (1961; reprinted London: Spokesman, 1976), p. 52.

12. Overton, *Mans Mortallitie*, p. 11.

13. Ibid., pp. 20–40.

14. Ibid., pp. 40, 48 and 41.

15. Ibid., p. 54. The final page of the first edition is taken up with biblical verses which were missed out of the original printing and 'belong unto the second Chapter', p. 57.

16. H. N. Brailsford makes only one mention of the text in his study of the Levellers, arguing that it was 'a little book' which merely repeated the General Baptist heresy of mortalism 'with the addition of confirmation drawn from biology.' Christopher Hill in his encyclopedic work *The World Turned Upside Down* emphasises Overton's interest in brushing 'the whole theological approach to politics aside', thus removing *Mans Mortallitie* from the political (and implicitly important) Leveller canon. Brailsford, *Levellers and the English Revolution*, pp. 51–2; Hill, *The World Turned Upside Down: Radical Ideas During the English Revolution* (London: Temple Smith, 1972), p. 133.

17. Harold Fisch, 'Introduction' to Fisch, ed., *Mans Mortalitie* (Liverpool: Liverpool University Press, 1968), p. xvi.

18. Christopher Hill, 'Why Bother About the Muggletonians?', in Christopher Hill, Barry Reay and William Lamont, *The World of the Muggletonians* (London: Temple Smith, 1983), p. 8.

19. Brailsford, *Levellers and the English Revolution*, p. 33. Following this lead William Lamont has worried that in recent histories the Levellers' 'political and social ideas have seemed more interesting than their religious ones.' Lamont, 'Pamphleteering, the Protestant Concensus and the English Revolution', in R. C. Richardson and G. M. Ridden, ed., *Freedom and the English Revolution: Essays in History and Literature* (Manchester: Manchester University Press, 1986), p. 83.

20. Tolmie, *Triumph of the Saints*, p. 73. Tolmie is adapting Patrick Collinson's phrase 'rustic Pelagianism', from *The Elizabethan Puritan Movement* (London: Jonathan Cape, 1967), p. 37.

21. Overton, *Mans Mortallitie*, p. 5.

22. [John Murton], *A Description of What God hath Predestinated Concerning Man* (1620), p. 143.

23. G. W. Bromiley, *Baptism and the Anglican Reformers* (London: Lutterworth Press, 1953), p. 101.

24. John Robinson, *A Defence of the Doctrine Propounded by the Synode at Dort: Against John Murton and his Associates* (1624), p. 185.

25. [Murton], *Description of What God hath Predestinated*, pp. 118–19.

26. This is a counter to the antinomian sense in which the always-alreadiness of salvation allowed for the abandonment of moral laws. See Tolmie, *Triumph of the Saints*, p. 73.

27. Tolmie, *Triumph of the Saints*, p. 82.

28. Thomas Edwards, *The Second Part of Gangraena* (1646), pp. 17–18.

29. Norman T. Burns, *Christian Mortalism from Tyndale to Milton* (Cambridge, Mass.: Harvard University Press, 1972), p. 10.

30. Henry [sic] Bullinger, *Of the Reasonable Soul of Man* (1577), cited in Burns, *Christian Mortalism*, pp. 15–16.

31. Overton, *Mans Mortallitie*, p. 12.

32. Nigel Smith emphasises the significance of Coke's ideas to John Lilburne: he calls the relationship 'a dependence', and reproduces the frontispiece of Lilburne's *The Triall of Lieut. Collonel John Lilburne* (1649) in which Lilburne poses with a copy of Coke's *Institutes*. Smith, *Literature and Revolution*, p. 134. I want to trace a link between Overton and Coke here.

33. Overton, *Mans Mortallitie*, p. 19.

34. Ibid., p. 14.

35. Godfrey Goodman, *The Creatures Praysing God: Or, the Religion of Dumbe Creatures* (1622), 'To The Reader', sig.A2ʳ and p. 5. Goodman includes the natural world generally under the term 'creature', but his reference to Noah's Arts (discussed later) hints that animals are the central references here.

36. Ibid., p. 2. For Herbert's view, see pp. 3–4 above.

37. Ibid., p. 29.

38. Ibid., p. 3.

39. 'Godfrey Goodman', *D.N.B.* (Oxford: Oxford University Press, 1917), Volume VIII, pp. 132–3.

40. Jean-Claude Schmitt, *The Holy Greyhound: Guinefort, healer of children since the thirteenth century*, translated by Martin Thom (Cambridge: Cambridge University Press, 1983), p. 136.

41. Goodman, *Creatures Praysing God*, 'To The Reader', sig.A2ᵛ.

42. Overton, *Mans Mortallitie*, p. 49.

43. John Calvin, *A Commentarie of John Caluine, Vpon the first booke of Moses called Genesis* (1578), p. 179.

44. Henry Vesey, *The Scope Of The Scripture* (1633), p. 8.

45. Thomas Draxe, *The Earnest of our Inheritance* (1613). For a discussion of Draxe see above, pp. 39–40.

46. Overton, *Mans Mortallitie*, p. 17. On the conventionality of this see Manfred Pfister, '"Man's Distinctive Mark": Paradoxical Distinctions Between Man and His Bestial Other in Early Modern Texts', in F. Lehmann and B. Lenz, ed., *Telling Stories: Studies in Honour of Ulrich Boich on the Occasion of his Sixtieth Birthday* (Amsterdam: B. R. Gruner, 1992), p. 23.

47. Overton, *Mans Mortallitie*, p. 17.

48. Ibid., p. 18.

49. Ibid., p. 7.

50. Ibid., p. 17.

51. This passage has been informed by the deconstructive theory of the supplement, for which see especially Jacques Derrida, '"... That Dangerous Supplement ..."', in Derrida, *Of Grammatology* (1967), translated by Gayatri Chakravorty Spivak (London: Johns Hopkins University Press, 1982), pp. 141–64, and Jonathan Culler, *On Deconstruction: Theory and Criticism after Structuralism* (1983; reprinted London: Routledge, 1994), p. 104.

52. Overton, *Mans Mortallitie*, p. 33.

53. See William Perkins, *An Exposition of the Symbole* (1595), in *The Workes of That Famovs and Worthy Minister of Christ in the Vniuersitie of Cambridge, Mr William Perkins* (1616–18), Volume I (1616).

54. William Perkins, *The Foundation of the Christian Religion Gathered into Six Principles* (1591), in *Workes*, Volume I, p. 4.

55. Thomas Dekker, *Worke for Armourours: Or, the Peace is Broken* (1609), sig.B2^r.

56. Gerrard Winstanley, *Light Shining in Buckinghamshire* (1648), p. 1.

57. Gerrard Winstanley, *The New Law of Righteousness* (1649), in Leonard Hamilton, ed., *Gerrard Winstanley: Selections from his Work* (London: Cresset Press, 1944), p. 18.

58. On the links between Overton and Winstanley see Maurice Goldsmith, 'Levelling by Sword, Spade and Word: Radical Egalitarianism in the English Revolution', in Colin James, Malyn Newitt and Stephen Roberts, ed., *Politics and People in Revolutionary England: Essays in Honor of Ivan Roots* (Oxford: Blackwell, 1986), p. 68.

59. *The Prerogative of Man: Or, His Soules Immortality* (1645), sig.B^v. Joseph Frank attributes this text to John Warre. Frank, *The Levellers: A History of the Writings of Three Seventeenth-Century Social Democrats: John Lilburne, Richard Overton, William Walwyn* (New York: Russell and Russell, 1955), p. 278.

60. *Prerogative of Man*, sigs.B^v and B2^r.

61. Thomas Edwards, *Gangraena* (1646), title page, and pp. 20, 26 and 27. Edwards also attacked what he perceived to be the levelling tendency of Leveller ideas. He writes of another heresy: 'That Pigeons in Dove Houses are common for all men to take and eat them, as well as those who are owners of those Dove Houses, because Pigeons are fowls of the aire, and so common to the sons of men.' (p. 9). This is a response to a legal debate about the ownership of pigeons such as that found in 'Dewell *versus* Sanders' (1619), recorded in Sir George

Croke, *The Second Part of the Reports of Sir George Croke* (1683), p. 492. For a summary of the changing place of the animal in response to the changing place of the land see P. B. Munsche, 'Introduction', in *Gentlemen and Poachers: The English Game Laws 1671–1831* (Cambridge: Cambridge University Press, 1981).

62. [Overton], *Arraignment of Mr Persecution*, p. 230.
63. Ibid., p. 236.
64. Edward Coke, *Quinta Pars Relationum Edwardi Coke* (1607), sig.Avi[r].
65. Richard Overton, *An Arrow Against All Tyrants and Tyrany* (1646), p. 3.
66. Ibid., p. 5.
67. C. B. Macpherson, *The Political Theory of Possessive Individualism: Hobbes to Locke* (Oxford: Oxford University Press, 1962), p. 107. On the limiting of the franchise see John Lilburne and others, *To The Supreme Authority of England, the Commons Assembled in Parliament* (1648), in Wolfe, ed., *Leveller Manifestoes*, especially p. 269. For a critique of Macpherson see Keith Thomas, 'The Levellers and the Franchise', in G. E. Aylmer, ed., *The Interregnum: The Quest for Settlement 1646–1660* (London: Macmillan, 1972), p. 59. The Levellers also imply that women be excluded as well. This is a point noted by R. C. Richardson and G. M. Ridden, 'Introduction', in Richardson and Ridden, ed., *Freedom and the English Revolution*, p. 11. For a discussion of the role of women in the Leveller movement see Ann Hughes, 'Gender and Politics in Leveller Literature', in Susan D. Amussen and Mark A. Kishlansky, ed., *Political Culture and Cultural Politics in Early Modern England: Essays Presented to David Underdown* (Manchester: Manchester University Press, 1995), pp. 162–88.
68. Richard Overton, *A Defiance Against All Arbitrary Usurpations* (1646), p. 2.
69. *A Dialogue Betwixt A Horse of Warre, and a Mill-Horse* (1645), sig.A3[r]. There is a slight echo in this pamphlet to the earlier horse-complaint by John Dando and Harry Runt, *Maroccus Extaticus: Or Bankes Bay Horse in a Trance* (1595).
70. Sir Philip Sidney, *The Old Arcadia* (1579), Katherine Duncan-Jones, ed. (Oxford: World's Classics, 1985), p. 224.
71. On this see Christopher Hill, *The Intellectual Origins of the English Revolution* (1965; reprinted London: Granada, 1972), p. 257; and Brian Manning, 'The Levellers and Religion', in J. F. McGregor and B. Reay, ed., *Radical Religion in the English Revolution* (Oxford: Oxford University Press, 1984), p. 67.
72. Gary K. Waite, 'Talking animals, preserved corpses and Venusberg: the sixteenth-century magical world view and popular conceptions of the spiritualist David Joris (c.1501–56)', *Social History*, 20: 2 (1995), 146 and 147.
73. George Wither, *A Collection of Emblems Ancient and Modern* (1635), p. 14.
74. Overton, *Mans Mortallitie*, pp. 17–18.
75. Overton, *Appeale From the degenerate Representative Body* (1647), in Wolfe, ed., *Leveller Manifestoes*, p. 158.

76. George Wither, 'Preface to the Reader', in Nemesius, *Of the Nature of Man, Englished by George Wither* (1636), sig.A4ᵛ.

77. Ibid., sigs.A4ᵛ–A5ʳ.

78. Brian Manning, *The English People and the English Revolution* (1976; reprinted Harmondsworth: Penguin, 1978), pp. 334–6.

79. Macpherson, *Political Theory of Possessive Individualism*, p. 145. Italics added.

80. A. L. Beier and Roger Finlay, 'Introduction: The Significance of the Metropolis', in Beier and Finlay, ed., *London 1500–1700: The Making of the Metropolis* (London: Longman, 1986), p. 20.

81. F. J. Fisher has emphasised the centrality of London to the English economy, a point repeated by Beier and Finlay. Fisher, 'London as an "Engine of Economic Growth"', in J. S. Bromiley and E. H. Kossmann, ed., *Britain and the Netherlands: Volume IV: Metropolis, Dominion and Province* (The Hague: Martinus Nijhoff, 1971), p. 3; Beier and Finlay, 'Introduction', p. 11.

82. Thomas, 'Levellers and the Franchise', pp. 71–2.

83. Christopher Hill, 'The Norman Yoke', in Hill, *Puritanism and Revolution: Studies in Interpretation of the English Revolution of the Seventeenth Century* (1958; reprinted London: Penguin, 1990), pp. 58–125.

84. See, for example, Lilburne and others, *To The Supream Authority of England*, pp. 266–7.

85. Hill, *Puritanism and Revolution*, p. 81.

86. This term comes from Douglas Duncan, *Ben Jonson and the Lucianic Tradition* (Cambridge: Cambridge University Press, 1979), p. 128. I use it earlier in relation to humanist ideas, p. 86.

87. J. C. Davis, 'Religion and the Struggle for Freedom in the English Revolution', *The Historical Journal*, 35: 3 (1992), 523.

88. Davis, 'Religion and the Struggle for Freedom', 521.

89. Lilburne and others, *To the Supream Authority of England*, p. 269.

90. [Overton], *Arraignment of Mr Persecution*, p. 236.

91. Overton, *An Appeale*, pp. 180–1. It is worth also noting that Overton proposes that the arbitrary rule of the House of Lords destroys the status of the rulers themselves. They 'esteeme soure sweet, and sweet soure'; they lose the ability to judge. Overton, *A Defiance*, p. 2.

92. Overton, *Mans Mortallitie*, p. 24.

Epilogue

1. Richard Overton, *Overton's Defyance of the Act of Pardon* (1649), p. 6.

2. Ibid., p. 4.

3. Richard Overton, *The Baiting of the Great Bull of Bashan unfolded* (1649), sig.A1ᵛ.

4. Overton, *Overton's Defyance*, p. 3.

5. Overton, *Baiting of the Great Bull*, sigs.A2ʳ and A1ᵛ.

6. Ibid., sig.A1ᵛ.

Notes

7. Sir Philip Sidney, *A Defence of Poetry*, (1579), J. A. Van Dorsten, ed. (Oxford: Oxford University Press, 1989), p. 53. Sidney is writing about reading Aesop's fables. I refer to this passage above in Ch. 3.
8. Overton, *Baiting of the Great Bull*, sig.A2[r].
9. Overton, *Overton's Defyance*, p. 6.
10. Ibid., p. 6.
11. Overton, *Overton's Defyance*, p. 4.
12. Overton, *Baiting of the Great Bull*, sig. A4[r].

Bibliography

UNPUBLISHED PRIMARY SOURCES

East Sussex Record Office, Lewes
    Q/R/E 58 – Q/R/E 82 (1642–9), Depositions
Public Record Office, London (Northern Circuit, Depositions concerning
    bestiality)
    ASSI 45/1/4, 17 (1642)
    ASSI 45/2/1, 259c (1647)
    ASSI 45/4/1, 134 (1651)
    ASSI 45/5/3, 25 and 26 (1656)
    ASSI 45/5/6, 8 (1659)
    ASSI 45/7/1, 138 (1664)
    ASSI 45/9/3, 85 (1670)
    ASSI 45/10/3, 210 (1673)
    ASSI 45/11/1, 165 and 166 (1674)
    ASSI 45/12/2, 90 (1675)
    ASSI 45/12/3, 106 (1679)
    ASSI 45/13/2, 1 and 2 (1682)
    ASSI 45/13/2, 20 and 21 (1682)
    ASSI 45/15/3, 106 (1689)

PUBLISHED PRIMARY SOURCES (PLACE OF PUBLICATION LONDON UNLESS OTHERWISE NOTED)

Adams, Thomas, *The Workes of Tho: Adams* (1630).
Albertus Magnus, *The Book of Secrets of Albertus Magnus*, Michael R. Best and
    Frank H. Brightman, ed. (Oxford: Clarendon, 1973).
—— *The Secrets of Albertvs Magnvs. Of the Vertues of Hearbes, Stones, and
    certaine Beasts* (1617).
Anon., *A Dialogue Betwixt a Horse of Warre, and a Mill-Horse* (1645).
Anon., *A true Discourse Declaring the damnable life and death of one Stubbe
    Peeter* (1590).
Anon., *The Booke of Common Prayer* (1564).
Anon., *The Fables of Esope in English* (c.1570).
Anon., *The History of Valentine and Orson* (c.1505).
Anon., *The Hystory of the two Valyaunt Brethren* (1565).
Anon., *The Morall Philosophie of Doni* (1570; 1601 edition).
Anon., *The Prerogative of Man: Or, His Soules Immortality* (1645).
Anon., *The Second Tome of Homilies* (1563).
Anon., *The three Wonders of this Age* (1636).
Anon., *Valentine and Orson. The Two Sonnes of the Emperour of Greece* (1637).

Anon., *Valentine and Orson*, Arthur Dickson ed. (EETS O.S. 204, 1937).

Aquinas, Thomas, *Summa Contra Gentiles*, translated by James F. Anderson (New York: Doubleday, 1956).

—— *Summa Theologiae*, translated by R. J. Batten (Blackfriars, 1975).

Arthington, Henry, *Principall Points of holy profession* (1607).

Babington, Gervase, *Certaine Plaine, briefe, and comfortable Notes vpon euerie Chapter of Genesis* (1592).

Bacon, Francis, *The Works of Francis Bacon*, James Spedding, Robert Leslie Ellis and Douglas Denon Heath ed. (1859; reprinted, Stuttgart: Friedrich Fromann, 1963).

—— *The Philosophy of Francis Bacon: An Essay on its Development from 1603 to 1609 with New Translations of Fundamental Texts*, Benjamin Farrington, ed. (Liverpool: Liverpool University Press, 1964).

Barron, Caroline, Christopher Coleman and Clare Gobbi, ed., 'The London Journal of Alessandro Magno 1562', *The London Journal*, 9 (1983), 136–52.

Batman, Stephan, *Batman vppon Bartholome, his Booke De Proprietatibus Rerum* (1582).

Beard, Thomas, *The Theatre of Gods Iudgements* (1597).

Boguet, Henri, *An Examen of Witches* (1590), translated by E. Allen Ashwin (Richard Clay & Sons, 1929).

Bolton, Robert, *Some Generall Directions For a Comfortable Walking with God* (1625).

Boyle, Robert, *The Works of the Honourable Robert Boyle in Six Volumes* (1772).

Breton, Nicholas, *The Court and the Country* (1618), in Alexander B Gossart, ed., *The Works in Verse and Prose of Nicholas Breton* (New York: A.M.S. Press, 1966), Volume II.

[Brinsley, John], *Esops Fables Translated both Grammatically, and also in propriety of our English phrase* (1624).

Bülow, Gottfried von, 'Diary of the Journey of Philip Julius, Duke of Stettin-Pomerania, through England in the Year 1602', *T.R.H.S.*, NS 6 (1892), 1–68.

Bulwer, John, *Anthropometamorphosis: Man Transform'd; Or, The Artificial Changeling* (1653).

Caius, John, *Of Englishe Dogges: The diversities, the names, the natures, and the Properties, translated by Abraham Fleming* (1576).

Calvin, John, *Institutes of the Christian Religion* (1559 edition), translated by Henry Beveridge (James Clarke & Co, 1949).

—— *A Commentarie of John Calvin, Vpon the first booke of Moses called Genesis* (1578).

Clarke, John, *Phraeologia puerilis Anglo-Latina* (1638).

Coke, Edward, *The First Part of the Institutes of the Lawes of England* (1628).

—— *Premier Pars de Les Reports* (1600; reprinted, 1609).

—— *Quinta Pars Relationum Edwardi Coke* (1605; reprinted, 1607).

—— *La Sept Part Des Reports* (1608).

—— *La Dix^{me} Part des Reports* (1614).

—— *The Third Part of the Institutes of the Laws of England* (1644).

—— *The Reports of Sir Edward Coke* (1658).

Connor, Bernard, *The History of Poland in Several Letters to Persons of Quality* (1698).

Cooke, Edward, *Bartas Junior. Or, The Worlds Epitome, Man* (1631).

Crab, Roger, *The English Hermite, Or Wonder of this Age* (1655).

—— *Dagons Down-fall; or, the great IDOL digged up Root and Branch* (1657).

Cressey, David, ed., *Education in Tudor and Stuart England* (Edward Arnold, 1975).

Croke, Sir George, *The Reports of Sir George Croke* (1658).

—— *The Second Part of the Reports of Sir George Croke* (1683).

Crowley, Robert, *A Briefe Discourse against the outwarde apparell and Ministring garmentes of the popishe church* (1566).

Dalton, Michael, *The Covntrey Jvstice* (1618).

Dando, John and Harry Runt, *Maroccus Extaticus. Or, Bankes Bay Horse in a Trance* (1595).

Dekker, Thomas, *The Seuen deadly Sinnes of London* (1606).

—— *Worke for Armourours: Or, The Peace is Broken* (1609).

Descartes, René, *Discourse on the Method* (1637), translated by F. E. Sutcliffe (Harmondsworth: Penguin, 1977).

Digby, Sir Kenelm, *Two Treatises: In the One of Which, The Nature of Bodies; in the Other The Nature of Man's Soule* (1644).

Dod, John and Robert Cleaver, *A Treatise or Exposition Vpon the Ten Commandments* (1603).

Draxe, Thomas, *The Earnest of our Inheritance* (1613).

Edwards, Thomas, *Gangraena* (1646).

—— *The Second Part of Gangraena* (1646).

Farley, Henry, *St. Pavles-Chvrch Her Bill for the Parliament* (1621).

Fenton, Edward, *Certaine Secrete wonders of Nature* (1569).

Fletcher, Joseph, *The Historie of the Perfect-Cursed Blessed Man* (1629).

Fox, George, *The Journal of George Fox*, John L. Nickalls, ed. (Religious Society of Friends, 1975).

Foxe, John, *A Sermon Preached at the Christening of a Iew, at London* (1578).

Fuller, Thomas, *The History of the Worthies of England* (1662; reprinted, 1811).

Gardiner, S. R., ed., *Documents Relating to the Proceedings Against William Prynne in 1634 and 1637*, Camden Society, n.s. 18 (1877).

Gifford, George, *A Dialogve concerning Witches and Witchecraftes* (1593).

Goodman, Godfrey, *The Creatures Praysing God: Or, the Religion of Dumbe Creatures* (1622).

Gouge, William, *Of Domesticall Duties* (1634).

H. W. [William Hubbocke], *An Apologie of Infants in a Sermon* (1595).

Hake, Edward, *Newes out of Powles Churchyarde* (1579).

Harvey, William, *An Anatomical Disputation Concerning the Movement of the Heart and Blood in Living Creatures* (1628), Gwenneth Whitteridge ed. (Oxford: Blackwell, 1976).

Herbert, George, *The English Poems of George Herbert*, C. A. Patrides, ed. (Dent, 1974).

Hill, Adam, *The Crie of England. A Sermon Preached at Paules Crosse* (1595).

Holinshed, Raphael, *Chronicles of England, Scotland, and Ireland* (1587).

Holland, Henry, *The Historie of Adam, or the foure-fold state of Man* (1606).

—— *A Treatise Against Witchcraft* (Cambridge, 1590).

Howarth, R. G., ed., *Minor Poets of the Seventeenth Century* (Dent, 1953).

James VI and I, *Daemonologie* (1597; reprinted 1605).

Johnson, Richard, *Looke on me London* (1613).

Jonson, Ben, *Ben Jonson*, C. H. Herford and Percy Simpson ed. (Oxford: Clarendon, 1927).

Junior, Democritus [Robert Burton], *The Anatomy of Melancholy* (1621; reprinted 1624).

[Le Loyer, Pierre], *A Treatise of Specters or straunge Sights, Visions and Apparitions appearing sensibly vnto men* (1605).

Lambard, William, *Eirenarcha: or of the Offices of Justices of the Peace in foure Bookes* (1592).

Leigh, Dorothy, *The Mothers Blessing* (1627).

Lenton, Francis, *The Young Gallants Whirligigg* (1629).

Lilburne, John and others, *To the Supream Authority of* England (1648), in Don M. Wolfe ed., *Leveller Manifestoes of the Puritan Revolution* (1944; reprinted Frank Cass, 1967).

Lupton, Donald, *London & The Country Carbonadoed and Quartered into Severall Characters* (1632).

Machiavelli, Niccolo, *The Letters of Machiavelli*, translated by Allan Gilbert (New York: Capricorn, 1961).

Markham, Gervase, *Countrey Contentments* (1615).

——— *The Whole Art of Husbandrie* (1631).

Marston, John, *The Scourge of Villanie* (1598), in Arnold Davenport, ed., *The Poems of John Marston* (Liverpool: Liverpool University Press, 1961).

Mascall, Leonard, *The first Booke of Cattell* (1591).

Microphilus, *The New-Yeeres Gift* (1636).

Moore, John, *A Mappe of Mans Mortalitie* (1617).

[Morton, Thomas], *A Treatise of the threefolde state of man* (1596).

Mulcaster, Richard, *Positions* (1581).

[Munday, Anthony], *A Second and Third Blast of Retrait from Plaies and Theaters* (1580).

[Murton, John], *A Description of What God hath Predestinated Concerning Man* (1620).

Nashe, Thomas, *Pierce Penniless* (1592), in J. B. Steane, ed., *The Fortunate Traveller and Other Works* (Penguin, 1985).

Nemesius, *Of the Nature of Man, Englished by George Wither* (1636).

Overton, Richard, *Mans Mortallitie* (1643/4).

——— *Mans Mortalitie* (1644), Harold Fisch, ed. (Liverpool: Liverpool University Press, 1968).

——— *The Arraignment of Mr Persecution* (1645), in William Haller, ed., *Tracts on Liberty in the Puritan Revolution* (Temple Smith, 1972).

——— *A Defiance Against All Arbitrary Usurpations* (1646).

——— *An Arrow Against All Tyrants and Tyranny* (1646).

——— *An Appeale From the degenerate Representative Body* (1647), in Don M. Wolfe, ed., *Leveller Manifestoes of the Puritan Revolution* (1944; reprinted Frank Cass, 1967).

——— *Overton's Defyance of the Act of Pardon* (1649).

——— *The Baiting of the Great Bull of Bashan unfolded* (1649).

——— *Man Wholly Mortal* (1655).

Paré, Ambroise, *Of Monsters and Prodigies* (1573), in *The Works of that Famous Chirurgeon Ambrose Parey* (1678).

Pasquil [Nicholas Breton?], *Pasquils Mad-cappe, Throwne at the Corruptions of these Times* (1626).

Perkins, William, *The Works of That Famous and Worthy Minister of Christ in the Vniuersitie of Cambridge, Mr William Perkins* (1616–18).

Pico della Mirandola, Giovanni, *On The Dignity of Man* (1486), translated by Charles Glenn Wallis (Indianapolis: Bobbs-Merrill, 1965).

Prynne, William, *Histrio-Mastix: The Players Scourge, or Actors Tragedie* (1633).

Puttenham, George, *The Arte of English Poesie* (1589), Gladys Doidge Willcock and Alice Walker, ed. (Cambridge: Cambridge University Press, 1936).

Rainoldes, John, *The overthrow of Stage-Playes* (1599; reprinted 1629).

Rankins, William, *A Mirrour of Monsters* (1587).

Rawlidge, Richard, *A Monster late found out and Discovered: or, the Scourging of Tiplers* (1628).

Rawlinson, John, *Mercy to a Beast. A Sermon Preached at Saint Maries Spittal in London on Tuesday in Easterweeke, 1612* (Oxford: 1612).

Reynolds, Henry, *Mythomystes Wherein A Short Survay is Taken of the Natvre and Valve of Trve Poesy* (1632).

Robinson, John, *A Defence of the Doctrine Propounded by the Synode at Dort: Against John Murton and his Associates* (1624).

Scot, Thomas, *Phylomythie or Phylomythologie wherin Outlandish Birds, Beasts, and Fishes are Taught to Speak true English plainely* (1616).

Scot, Reginald, *The discouerie of witchcraft* (1584).

Shakespeare, William, *William Shakespeare: The Complete Works*, Stanley Wells and Gary Taylor, ed. (Oxford: Clarendon, 1988).

Sidney, Sir Philip, *A Defence of Poetry* (1579), J. A. Van Dorsten, ed. (Oxford: Oxford University Press, 1989).

—— *The Old Arcadia* (1579), Kathernine Duncan-Jones, ed. (Oxford: World's Classics, 1985).

—— *A Discourse of Syr Ph. S. To The Queenes Majesty Touching Hir Mariage With Monsieur* (1579), in Albert Feuillerat, ed., *The Prose Works of Sir Philip Sidney* (Cambridge: Cambridge University Press, 1963), Volume III.

—— *Astrophil and Stella* (1591), in Katherine Duncan-Jones, ed., *Sir Philip Sidney: Selected Poems* (Oxford: Clarendon, 1988).

Slotkin, J. S., ed., *Readings in Early Anthropology* (Chicago: Aldine, 1965).

Spenser, Edmund, *The Shepheardes Calender* (1579), in J. C. Smith and E. de Selincourt, ed., *The Poetical Works of Edmund Spenser* (Oxford: Oxford University Press, 1924).

—— *A viewe of the presente state of Irelande* (1596), in Rudolf Gottfried, ed., *Spenser's Prose Works* (Baltimore: Johns Hopkins University Press, 1949).

Sprat, Thomas, *The History of the Royal Society* (1667).

Stubbes, Phillip, *The Anatomie of Abuses* (1583).

Taylor, John, The Water Poet, *Wit and Mirth* (1630), in W. Carew Hazlitt, ed., *Shakespeare Jest Books* (Willis and Southeran, 1864).

—— *Bull, Beare, and Horse* (1638), in *The Works of John Taylor The Water Poet Not Included in the Folio Volume of 1630* (Manchester: Spenser Society, 1876).

Taylor, Thomas, *A Vindication of the Rights of Brutes* (1792).

Topsell, Edward, *Times Lamentation* (1599).

—— *The Historie of Foure-Footed Beastes* (1607).
Turner, William, *A Compleat History of the Most Remarkable Providences* (1697).
Vale, Marcia ed., *The Gentleman's Recreations: Accomplishments and Pastimes of the English Gentleman 1580–1630* (Cambridge: D. S. Brewer, 1977).
Vaughan, William, *The Golden Grove* (1626).
Vesey, Henry, *The Scope Of The Scripture* (1633).
Webster, John, *The Duchess of Malfi* (1613), in Jonathan Dollimore and Alan Sinfield, ed., *The Selected Plays of John Webster* (Cambridge: Cambridge University Press, 1983).
White, T. H., ed., *The Book of Beasts: Being a Translation from a Latin Bestiary of the Twelfth Century* (1954; reprinted, Stroud: Allan Sutton, 1992).
Whitney, Geffrey, *A Choice of Emblemes and Other Devices* (1586).
Wilson, George, *The Commendation of Cockes and Cock-fighting* (1607).
Winstanley, Gerrard, *Light Shining in Buckinghamshire* (1648).
—— *The New Law of Righteousness* (1649), in Leonard Hamilton, ed., *Gerrard Winstanley: Selections from his Work* (Cresset Press, 1944).
Wither, George, *A Collection of Emblems Ancient and Modern* (1635).
Wright, Thomas, ed., *Alexandri Neckam: De Naturis Rerum* (Longman, 1863).

SECONDARY SOURCES.

Altick, Robert D., *The Shows of London* (London: Belknap Press, 1978).
Baker, J. H., *The Legal Profession and the Common Law* (London: Hambledon, 1986).
—— *An Introduction to English Legal History*, third edition (London: Butterworths, 1990).
Beier, A. L., *Masterless Men: The Vagrancy Problem in England 1560–1660* (London: Methuen, 1985).
—— and Roger Finlay, ed., *London: 1500–1700: The Making of the Metropolis* (London: Longman, 1986).
Beilin, Elaine, *Redeeming Eve: Women Writers of the English Renaissance* (Princeton N.J.: Princeton University Press, 1987).
Bentley, G. E., *The Jacobean and Caroline Stage* (Oxford: Clarendon, 1968).
Bernheimer, Richard, *Wild Men In The Middle Ages: A Study in Art, Sentiment and Demonology* (Cambridge, Mass.: Harvard University Press, 1952).
Bhabha, Homi, 'Of Mimicry and Man: The Ambivalence of Colonial Discourse', *October*, 28 (1984), 125–33.
Billington, Sandra, 'Butchers and Fishmongers: Their Historical Contributions to London's Festivity', *Folklore*, 101: 1 (1990), 97–103.
Boughner, Daniel C., *The Devil's Disciple: Ben Jonson's Debt to Machiavelli* (New York: Philosophical Library, 1968).
Bowden, Peter J., *The Wool Trade in Tudor and Stuart England* (London: Macmillan, 1962).
Bowen, Margarita, *Empiricism and Geographical Thought: From Francis Bacon to Alexander von Humboldt* (Cambridge: Cambridge University Press, 1981).
Brailsford, Dennis, *Sport in Society: Elizabeth to Anne* (London: Routledge, 1969).

Brailsford, H. N., *The Levellers and the English Revolution*, Christopher Hill, ed. (1961; reprinted London: Spokesman, 1976).

Breward, Ian, 'Introduction' to Breward, ed., *The Work of William Perkins* (Appleforth: Sutton Courtenay Press, 1970).

Briggs, William Dinsmore, 'Political Ideas in Sidney's Arcadia', *Studies in Philology*, 28:2 (1931), 137–61.

Bromiley, G. W., *Baptism and the Anglican Reformers* (London: Lutterworth Press, 1953).

Brown, Les, *Cruelty to Animals: The Moral Debt* (London: Macmillan, 1988).

Brownstein, Oscar, 'The Popularity of Baiting before 1600: A Study in Social and Theatrical History', *Educational Theatre Journal*, 21 (1969), 237–50.

Burke, Peter, 'Popular Culture in Seventeenth-Century London', *The London Journal*, 3: 2 (1977), 143–62.

Burns, Norman T., *Christian Mortalism from Tyndale to Milton* (Cambridge, Mass.: Harvard University Press, 1972).

Canup, John, '"The Cry of Sodom Enquired Into": Bestiality and the Wilderness of Human Nature in Seventeenth-Century New England', *Proceedings of the American Antiquarian Society*, 98 (1988), 113–34.

Carlin, Norah, 'Ireland and Natural Man in 1649', in Francis Barker *et al.*, ed., *Europe and its Others* (Colchester: University of Essex Press, 1985), pp. 91–111.

Carroll, William Meredith, *Animal Conventions in English Renaissance Non-Religious Prose (1550–1600)* (New York: Bookman Associates, 1954).

Chambers, E. K., *The Elizabethan Stage* (Oxford: Clarendon, 1923).

Chartier, Roger, *The Order of Books: Readers, Authors and Libraries in Europe Between the Fourteenth and the Eighteenth Centuries*, translated by Lydia G. Cochrane (Cambridge: Polity Press, 1994).

Clark, Peter, *The English Ale House: A Social History* (London: Longman, 1983).

Clark, Peter and David Souden ed., *Migration and Society in Early Modern England* (London: Hutchinson, 1987).

Cohen, Esther, 'Law, Folklore and Animal Lore', *Past and Present*, 110 (1986), 6–37.

Cole, F. J., 'Harvey's Animals', *Journal of the History of Medicine*, 12 (1957), 106–113.

Collinson, Patrick, *The Elizabethan Puritan Movement* (London: Jonathan Cape, 1967).

Culler, Jonathan, *On Deconstruction: Theory and Criticism After Poststructuralism* (1983; reprinted, London: Routledge, 1994).

Davies, Stevie, 'Introduction', to Davies, ed., *Renaissance Views of Man* (Manchester: Manchester University Press, 1978).

Davies, Tony, *Humanism* (London: Routledge, 1997).

Davis, David Brion, *The Problem of Slavery in Western Culture* (Ithaca: Cornell University Press, 1966).

Davis, J. C., 'The Levellers and Christianity', in Brian Manning, ed., *Politics, Religion and the English Civil War* (London: Edward Arnold, 1973), pp. 224–59.

—— 'Religion and the Struggle for Freedom in the English Revolution', *The Historical Journal*, 35: 3 (1992), 507–30.

Dawson, Giles E., 'London's Bull-baiting and Bear-baiting Arena in 1562', *Shakespeare Quarterly*, 15 (1964), 97–101.

Debus, Allen, *Man and Nature in the Renaissance* (Cambridge: Cambridge University Press, 1978).

De Mause, Lloyd, 'The Evolution of Childhood', in De Mause, ed., *The History of Childhood: The Untold Story of Child Abuse*, second edition (New York: Peter Bedrick Books, 1988), pp. 1–73.

Derrida, Jacques, *Of Grammatology* (1967), translated by Gayatri Chakravorty Spivak (Baltimore: Johns Hopkins University Press, 1974).

Donaldson, Ian, 'Jonson's Tortoise', in Jonas A. Barish, ed., *Volpone: A Casebook* (London: Macmillan, 1972) , pp. 189–94.

Douglas, Adam, *The Beast Within: Man, Myths and Werewolves* (London: Orion, 1992).

Drum, Peter, 'Aquinas and the Moral Status of Animals', *American Catholic Philosophical Quarterly*, 66: 4 (1992), 483–8.

Dubois-Desaulle, Gaston, *Bestiality: An Historical, Medical, Legal and Literary Study* (c.1905), translated by A. F. N. (New York: Panurge Press, 1933).

Dudley, Edward and Maximillian E. Novak, ed., *The Wild Man Within: An Image in Western Thought from the Renaissance to Romanticism* (London: University of Pittsburg Press, 1972).

Duncan, Douglas, *Ben Jonson and the Lucianic Tradition* (Cambridge: Cambridge University Press, 1979).

Durston, Christopher and Jacqueline Eales, ed., *The Culture of English Puritanism, 1560–1700* (Basingstoke: Macmillan, 1996).

Eco, Umberto, 'The frames of comic freedom', in Thomas A. Sebeok, ed., *Carnival!* (Berlin: Mouton, 1984), pp.1–9.

Emmison, F. G., *Elizabethan Life: Disorder* (Chelmsford: Essex County Council, 1970).

Evans, E. P., *The Criminal Prosecution and Capital Punishment of Animals: The Lost History of Europe's Animal Trials* (1906; reprinted London: Faber and Faber, 1988).

Fairholt, E. W., *Remarkable and Eccentric Characters* (London: Richard Bentley, 1849).

Feinberg, Anat, '"Like Demie Gods the Apes Began to Move": The Ape in the English Theatrical Tradition, 1580–1660', *Cahiers Elisabethains: Sur la Pre Renaissance et la Renaissance Anglaises*, 35 (1989), 1–13.

Fiddes, Nick, *Meat: A Natural Symbol* (London: Routledge, 1991).

Finkelstein, J. J., 'The Ox That Gored', *Transactions of the American Philosophical Society*, 7: 2 (1981).

Fisher, F. J., 'The Development of London as a Centre of Conspicuous Consumption in the Sixteenth and Seventeenth Centuries', *T.R.H.S.*, 4th Series, 3 (1948), 37–50.

—— 'London as an "Engine of Economic Growth"', in J. S. Bromiley and E. H. Kossmann, ed., *London and the Netherlands Volume IV: Metropolis, Dominion and Province* (The Hague: Martinus Nijhoff, 1971), pp. 3–16.

Frank, Joseph, *The Levellers: A History of the Writings of Three Seventeenth-Century Social Democrats: John Lilburne, Richard Overton, William Walwyn* (New York: Russell and Russell, 1955).

Freeman, Rosemary, *The English Emblem Books* (London: Chatto and Windus, 1948).

Freidman, John Block, *The Monstrous Races in Medieval Art and Thought* (London: Harvard University Press, 1981).

French, Richard D., 'Animal Experimentation: Historical Aspects', in Warren T. Reich, ed., *Encyclopedia of Bioethics* (New York: Free Press, 1978), pp. 75–9.

Fudge, Erica, 'Temples of God: William Prynne and the New Science', in Tracey Hill and Jeffrey Rodman, ed., *The Body of Truth: Corporeality and Power in Early Modern Culture* (Bath: Sulis Press, 1999).

Fudge, Erica, Ruth Gilbert, and Susan Wiseman, ed., *At The Borders of the Human: Beasts, Bodies and Natural Philosophy* (Basingstoke: Macmillan, 1999).

Fuss, Diana, ed., *Human, All too Human* (London: Routledge, 1996).

General Synod of the Church of England – 'What is the Church's View?' (Board for Social Responsibility, 1994).

George, Wilma and Brunsdon Yapp, *The Naming of the Beasts: Natural history in the medieval bestiary* (London: Duckworth, 1991).

Goldberg, Jonathan, *Sodomitries: Renaissance Texts, Modern Sexualities* (Stanford: Stanford University Press, 1992).

Goldsmith, Maurice, 'Levelling by Sword, Spade and Word: Radical Egalitarianism in the English Revolution', in Colin James, Malyn Newitt and Stephen Roberts, ed., *Politics and People in Revolutionary England: Essays in Honor of Ivan Roots* (Oxford: Blackwell, 1986), pp. 65–80.

Green, Ian, '"For Children in Yeeres and Children in Understanding": The Emergence of the English Catechism under Elizabeth and the Early Stuarts', *The Journal of Ecclesiastical History*, 37: 3 (1986), 397–425.

Greenblatt, Stephen, 'The False Ending in *Volpone*', *Journal of English and Germanic Philology*, 75 (1976), 90–104.

—— *Shakespearean Negotiations: The Circulation of Social Energy in Renaissance England* (Oxford: Clarendon, 1988).

—— *Marvellous Possessions: The Wonder of the New World* (Oxford: Clarendon, 1991).

Greven, Philip, *The Protestant Temperament: Patterns of Child Rearing, Religious Experience and the Self in Early America* (New York: Meridian, 1979).

Guerrini, Anita, 'The Ethics of Animal Experimentation in Seventeenth-Century England', *Journal of the History of Ideas*, 50: 3 (1989), 391–407.

Gurr, Andrew, *Playgoing in Shakespeare's London* (Cambridge: Cambridge University Press, 1987).

Ham, Jennifer and Senior, Matthew, ed., *Animals Acts: Configuring the Human in Western History* (London: Routledge, 1997).

Harding, Alan, *A Social History of the English Law* (Harmondsworth: Penguin, 1966).

Harrison, Peter, 'Animal Souls, Metempsychosis, and Theodicy in Seventeenth-Century English Thought', *Journal of the History of Philosophy*, 31: 4 (1993), 519–44.

Hawkes, David, *Ideology* (London: Routledge, 1996).

Heal, Felicity, 'The Idea of Hospitality in Early Modern England', *Past and Present*, 102 (1984), 66–93.

Heinemann, Margot, *Puritanism and Theatre: Thomas Middleton and Opposition Drama under the Early Stuarts* (Cambridge: Cambridge University Press, 1980).

Helgerson, Richard, *Forms of Nationhood: The Elizabethan Writing of England* (Chicago: University of Chicago Press, 1992).

Henricks, Thomas S., *Disputed Pleasures: Sport and Society in Pre-Industrial England* (Westport, Conn.: Greenwood Press, 1991).

Herrup, Cynthia B., *The Common Peace: Participation in the Common Law in Seventeenth-Century England* (Cambridge: Cambridge University Press, 1987).

Hill, Christopher, *Puritanism and Revolution: Studies in the Interpretation of the Seventeenth Century* (1958; reprinted, London: Penguin, 1990).

—— *The Intellectual Origins of the English Revolution* (1965; reprinted, London: Granada, 1972).

—— *The World Turned Upside Down: Radical Ideas During the English Revolution* (London: Temple Smith, 1972).

—— 'Why Bother About the Muggletonians?', in Christopher Hill, Barry Reay and William Lamont, *The World of the Muggletonians* (London: Temple Smith, 1983).

Hill, Tracey, '"He hath changed his coppy": Anti-Theatrical Writing and the Turncoat Player', *Critical Survey*, 9: 3 (1997), 59–77.

Hodgen, Margaret T., *Early Anthropology in the Sixteenth and Seventeenth Centuries* (Philadelphia: University of Pennsylvania Press, 1974).

Holdsworth, W. S., *A History of the English Law* (London: Methuen, 1924).

Holt, Richard, *Sport and the British: A Modern History* (Oxford: Clarendon, 1989).

Hotson, Leslie, *The Commonwealth and Restoration Stage* (Cambridge, Mass.: Harvard University Press, 1928).

Hughes, Ann, 'Gender and Politics in Leveller Literature', in Susan D. Amussen and Mark A. Kishlansky, ed., *Political culture and cultural politics in early modern England: Essays presented to David Underdown* (Manchester: Manchester University Press, 1995), pp. 162–88.

Hulme, Peter, *Colonial Encounters: Europe and the Native Carribean, 1492–1797* (London: Methuen, 1986).

Hyde, W. W., 'The Prosecution and Punishment of Animals and Lifeless Things in the Middle Ages and Modern Times', *University of Pennsylvania State Law Review*, 64 (1916), 696–730.

Jacquart, Danielle and Claude Thomasset, *Sexuality and Medicine in the Middle Ages* (Cambridge: Polity Press, 1988).

Janson, H. W., *Apes and Ape Lore in the Middle Ages and the Renaissance* (London: Warburg Institute, 1952).

Jordan, Winthrop D., *White Over Black: American Attitudes Toward the Negro, 1550–1812* (Chapel Hill: University of North Carolina Press, 1968).

Joseph, Sister Miriam, *Shakespeare's Use of the Arts of Language* (New York: Columbia University Press, 1947).

Kendall, R. T., *Calvin and English Calvinism to 1649* (Oxford: Oxford University Press, 1979).

Kete, Kathleen, *The Beast in the Boudoir: Petkeeping in Nineteenth-Century Paris* (London: University of California Press, 1994).

Kingsford, C. L., 'Paris Garden and the Bear-Baiting', *Archaeologia*, 20 (1920), 155–78.

Klingender, Francis, *Animals in Art and Thought at the End of the Middle Ages* (London: Routledge, 1971).

Klug, Brian, 'Lab Animals, Francis Bacon and the Culture of Science', *Listening*, 18 (1983), 54–72.

Knappen, M. M., *Tudor Puritanism: A Chapter in the History of Idealism* (1939; reprinted London: Chicago University Press, 1970).

Kocher, Paul H., 'Francis Bacon on the Science of Jurisprudence', *Journal of the History of Ideas*, 18: 1 (1957), 3–26.

Lamont, William, *Marginal Prynne 1600–1669* (London: Routledge, 1963).

Lansbury, Coral, *The Old Brown Dog: Women, Workers and Vivisection in Edwardian England* (London: University of Wisconsin Press, 1985).

Leach, Edmund, 'Anthropological Aspects of Language: Animal Categories and Verbal Abuse', in Eric H. Lenneberg, ed., *New Directions in the Study of Language* (Cambridge. Mass.: MIT Press, 1966), pp. 23–63.

Levao, Ronald, 'Francis Bacon and the Mobility of Science', *Representations*, 40 (1992), 1–32.

Levinas, Emmanuel, 'The Name of a Dog, or Natural Rights', in *Difficult Freedom: Essays on Judaism*, translated by Séan Hand (London: Athlone Press, 1990), pp. 151–3.

Levine, Laura, 'Men in Women's Clothing: Anti-theatricality and Effeminization from 1579 to 1642', *Criticism*, 28: 2 (1986), 121–43.

—— *Men in Women's Clothing: Anti-theatricality and effeminization from 1579 to 1642* (Cambridge: Cambridge University Press, 1994).

Levy, Ellen K., and David E. Levy, 'Monkey in the Middle: Pre-Darwinian Evolutionary Thought and Artistic Creation', *Perspectives in Biology and Medicine*, 30 (1986), 95–106.

Lewis, John Underwood, 'Sir Edward Coke (1552–1633): His Theory of "Artificial Reason" As a Context for Modern Basic Legal Theory', *Law Quarterly Review*, 84: 335 (1968), 330–42.

Lovejoy, Arthur O., *The Great Chain of Being: A Study of the History of an Idea* (London: Harvard University Press, 1936).

MacDonald, Michael, *Mystical Bedlam: Madness, Anxiety and Healing in Seventeenth-Century England* (Cambridge: Cambridge University Press, 1981).

Macpherson, C. B., *The Political Theory of Possessive Individualism: Hobbes to Locke* (Oxford: Oxford University Press, 1962).

Maehle, Andreas Holger and Ulrich Tröhler, 'Animal Experimentation from Antiquity to the end of the Eighteenth Century: Attitudes and Arguments', in N. A. Rupke, ed., *Vivisection in Historical Perspective* (London: Routledge, 1990), pp. 14–47.

Malson, Lucien, *Wolf Children*, translated by Edmund Fawcett, Peter Ayrton and Joan White (London: NLB, 1972).

Manning, Brian, *The English People and the English Revolution* (1976; reprinted Harmondsworth: Penguin, 1978).

—— 'The Levellers and Religion', in J. F. McGregor and B. Reay, ed., *Radical Religion in the English Revolution* (Oxford: Oxford University Press, 1984), pp. 65–90.

Marcus, Leah Sinanoglou, *Childhood and Cultural Despair: A Theme and Variations in Seventeenth-Century Literature* (Pittsburg: University of Pittsburg Press, 1978).

Marin, Louis, 'The "Aesop" Fable Animal', in Marshall Blonsky, ed., *On Signs* (Oxford: Blackwell, 1985), pp. 334–40.

Martin, Julian, *Francis Bacon, The State and the Reform of Natural Philosophy* (Cambridge: Cambridge University Press, 1992).

Martindale, Joanna, *English Humanism: Wyatt to Cowley* (Beckenham: Croom Helm, 1985).

Marx, Karl, *Karl Marx: Selected Writings*, David McLellan, ed. (Oxford: Oxford University Press, 1977).

Maslen, R. W., *Elizabethan Fictions: Espionage, Counter-Espionage and the Duplicity of Fiction in Early Elizabethan Prose Narratives* (Oxford: Clarendon, 1997).

Mason, Peter, 'The excommunication of caterpillars: ethno-anthropological remarks on the trial and punishment of animals', *Social Science Information*, 27: 2 (1988), 265–73.

Mastromarino, Mark A., 'Teaching Old Dogs New Tricks: The English Mastiff and the Anglo-American Experience', *The Historian*, 49 (1986), 10–25.

Mazzio, Carla, 'Sins of the Tongue', in Carla Mazzio and David Hillman, ed., *The Body in Parts: Fantasies of Corporeality in Early Modern England* (London: Routledge, 1997), pp. 53–79.

McGrath, Alister E., *The Intellectual Origins of the European Reformation* (Oxford: Blackwell, 1987).

—— *Reformation Thought: An Introduction*, second edition (Oxford: Blackwell, 1993).

McLean, Antonia, *Humanism and the Rise of Science in Tudor England* (London: Heinemann, 1972).

McPherson, David, 'Ben Jonson's Library and Marginalia: An Annotated Catalogue', *Studies in Philology*, 71: 5 (1974).

Merchant, Carolyn, *The Death of Nature: Women, Ecology and the Scientific Revolution* (1980; reprinted, London: Harper Collins, 1990).

Milsom, S. F. C., *Historical Foundations of the Common Law*, second edition (London: Butterworths, 1981).

Morgan, John, *Godly Learning: Puritan Attitudes Towards Reason, Learning and Education, 1560–1640* (Cambridge: Cambridge University Press, 1986).

Munsche, P. B., *Gentlemen and Poachers: The English Game Laws 1671–1831* (Cambridge: Cambridge University Press, 1981).

Needler, Howard, 'The Animal Fable Among Other Medieval Literary Genres', *New Literary History*, 22: 2 (1991), 423–39.

Nicolson, Marjorie, 'The Early Stage of Cartesianism in England', *Studies in Philology*, 26 (1929), 356–74.

Nietzsche, Friedrich, *The Gay Science* (1882), translated by Walter Kaufmann (New York: Vintage, 1974).

Norbrook, David, *Poetry and Politics in the English Renaissance* (London: Routledge and Keegan Paul, 1984).

Oaks, Robert F., '"Things Fearful to Name": Sodomy and Buggery in Seventeenth-Century New England', *Journal of Social History*, 12: 2 (1978), 268–81.

Oates, Caroline, 'Metamorphosis and Lycanthropy in Franche-Comté, 1521–1643', in Michel Feher with Ramona Nadodd and Nadia Tazi, ed., *Fragments for a History of the Human Body: Part One* (New York: Zone, 1989), pp. 305–63.

O'Connell, Michael, 'The Idolatrous Eye: Iconoclasm, Anti-theatricalism, and the Image of the Elizabethan Theater', *ELH*, 52: 2 (1985), 279–310.

Oster, Malcolm R., 'The "Beame of Diuinity": Animal Suffering in the Early Thought of Robert Boyle', *British Journal for the History of Science*, 22: 2 (1989), 151–79.

Pagden, Anthony, *The Fall of Natural Man: The American Indian and the Origins of Comparative Ethnology* (Cambridge: Cambridge University Press, 1982).

Park, Katherine and Lorraine Daston, 'Unnatural Conceptions: The Study of Monsters in Sixteenth- and Seventeenth-Century France and England', *Past and Present*, 92 (1981), 20–54.

Parker, R. B., 'Volpone and Reynard the Fox', *Renaissance Drama*, n.s. 7 (1976), 3–42.

Passmore, John, 'The Treatment of Animals', *Journal of the History of Ideas*, 36: 2 (1975), 195–218.

Paterson, Timothy H., 'Bacon's Myth of Orpheus: Power as a Goal of Science in *Of the Wisdom of the Ancients*', *Interpretation*, 16: 3 (1989), 427–44.

Patterson, Annabel, *Censorship and Interpretation: The Conditions of Writing and Reading in Early Modern England* (London: University of Wisconsin Press, 1984).

—— *Fables of Power: Aesopian Writing and Political History* (London: Duke University Press, 1991).

Pfister, Manfred, '"Man's Distinctive Mark": Paradoxical Distinctions Between Man and His Bestial Other in Early Modern Texts', in F. Lehmann and B. Lenz, ed., *Telling Stories: Studies in Honour of Ulrich Boich on the Occasion of his Sixtieth Birthday* (Amsterdam: B. R. Gruner, 1992), pp. 17–33.

Platt, Anthony Michael, and Bernard L. Diamond, 'The Origins and Development of the "Wild Beast" Concept of Mental Illness and Its Relation to Theories of Criminal Responsibility', *Journal of the History of the Behavioral Sciences*, 1: 4 (1965), 355–67.

Porter, H. C., *Reformation and Reaction in Tudor Cambridge* (Cambridge: Cambridge University Press, 1958).

Porter, Roy, ed., *Rewriting the Self: Histories from the Renaissance to the Present* (London: Routledge, 1997).

Porter, Roy and Mikuláš Teich, ed., *The Scientific Revolution in National Context* (Cmbridge: Cambridge University Press, 1992).

Pumfrey, Stephen, Paolo L. Rossi and Maurice Slawinski ed., *Science, culture and popular belief in Renaissance Europe* (Manchester: Manchester University Press, 1991).

Quinton, Anthony, *Francis Bacon* (Oxford: Oxford University Press, 1980).

Raab, Felix, *The English Face of Machiavelli: A Changing Interpretation 1500–1700* (London: Routledge, 1964).

Rambuss, Richard, *Spenser's Secret Career* (Cambridge: Cambridge University Press, 1993).

Raven, Charles E., *English Naturalists From Neckam to Ray: A Study in the Making of the Modern World* (Cambridge: Cambridge University Press, 1947).

Redwood, B. C., ed., *Quarter Sessions Order Book, 1642–1649* (Lewes: Sussex Record Society, 1954).

Richardson, R. C. and G. M. Ridden, ed., *Freedom and the English Revolution: Essays in History and Literature* (Manchester: Manchester University Press, 1986).

Ritvo, Harriet, *The Animal Estate: The English and Other Creatures in the Victorian Age* (London: Penguin, 1990).

—— *The Platypus and the Mermaid and Other Figments of the Classifying Imagination* (Cambridge, Mass.: Harvard University Press, 1997).

Rossi, Paolo, *Francis Bacon: From Magic to Science*, translated by Sasha Rabinovitch (London: Routledge, 1968).

Rowland, Beryl, *Animals with Human Faces: A Guide to Animal Symbolism* (London: Allen & Unwin, 1974).

Salisbury, Joyce E., *The Beast Within: Animals in The Middle Ages* (London: Routledge, 1994).

Sawday, Jonathan, 'The Fate of Marsyas: Dissecting the Renaissance Body', in Lucy Gent and Nigel Llewellyn, ed., *Renaissance Bodies* (London: Reaktion, 1990), pp. 111–35.

—— *The Body Emblazoned: Dissection and the Human Body in Renaissance Culture* (London: Routledge, 1995).

Scheve, D. A., 'Jonson's *Volpone* and Traditional Fox Lore', *Review of English Studies*, n.s. 1 (1950), 242–4.

Schmitt, Jean-Claude, *The Holy Greyhound: Guinefort, healer of children since the thirteenth century*, translated by Martin Thom (Cambridge: Cambridge University Press, 1983).

Schnucker, Robert V., 'Puritan attitudes towards childhood discipline, 1560–1634', in Valerie Fildes, ed., *Women as Mothers in Pre-Industrial England, Essays in Memory of Dorothy McLaren* (London: Routledge, 1990), pp. 108–21.

Serpell, James, *In the Company of Animals: A Study of Human-Animal Relationships* (Oxford: Blackwell, 1986).

Sharpe, J. A., *Crime in Early Modern England 1550–1750* (London: Longman, 1984).

Shell, Marc, 'The Family Pet', *Representations*, 15 (1986), 121–53.

Shugg, Wallace, 'Humanitarian Attitudes in the Early Animal Experiments of the Royal Society', *Annals of Science*, 24 (1968), 227–38.

Simons, John, 'The Longest Revolution: Cultural Studies After Speciesism', *Environmental Values*, 6 (1997), 483–97.

Sinfield, Alan, 'The Cultural Politics of *The Defence of Poetry*', in Gary F. Waller and Michael D. Moore, ed., *Sir Philip Sidney and the Reinterpretation of Renaissance Culture* (London: Croom Helm, 1984), pp. 124–43.

Smith, Nigel, *Literature and Revolution in England 1640–1660* (London: Yale University Press, 1994).

Sommerville, C. John, *The Discovery of Childhood in Puritan England* (London: University of Georgia Press, 1992).

South, Malcolm H., 'Animal Imagery in *Volpone*', *Tennessee Studies in Literature*, 10 (1965), 141–50.

Speigel, Marjorie, *The Dreaded Comparison: Human and Animal Slavery* (London: Heretic Books, 1988).

Spencer, Colin, *The Heretic's Feast: A History of Vegetarianism* (London: Fourth Estate, 1993).

Spierenberg, Pieter, *The Spectacle of Suffering: Executions and the Evolution of Repression: From a Preindustrial Metropolis to the European Experience* (Cambridge: Cambridge University Press, 1984).

Stachniewski, John, *The Persecutory Imagination: English Puritanism and the Literature of Religious Despair* (Oxford: Clarendon, 1991).

Stallybrass, Peter, and Allon White, *The Politics and Poetics of Transgression* (Ithaca: Cornell University Press, 1986).

Stewart, Alan, *Close Readers: Humanism and Sodomy in Early Modern England* (Princeton N.J.: Princeton University Press, 1997).

Symonds, W., 'Winterslow Church Reckonings, 1542–1660', *Wiltshire Archeological Magazine*, 36 (1909–10), 27–49.

Thomas, Keith, *Religion and the Decline of Magic: Studies in Popular Beliefs in Sixteenth- and Seventeenth-Century England* (1971; reprinted London: Penguin, 1991).

—— 'The Levellers and the Franchise', in G. E. Aylmer, ed., *The Interregum: The Quest for Settlement 1646–1660* (London: Macmillan, 1972), pp. 57–78.

—— *Man and the Natural World: Changing Attitudes in England 1500–1800* (London: Penguin, 1984).

Thompson, E. P., 'Patrician Society, Plebian Culture', *Journal of Social History*, 7 (1974), 382–405.

Tolmie, Murray, *The Triumph of the Saints: The Separate Churches of London 1616–1649* (Cambridge: Cambridge University Press, 1977).

Trigg, Jonathan D., *Baptism in the Theology of Martin Luther* (Leiden: E. J. Brill, 1994).

Trousdale, Margaret, *Shakespeare and the Rhetoricians* (London: Scolar Press, 1982).

Turner, E. S., *All Heaven in a Rage* (1964; reprinted, Fontwell: Centaur, 1992).

Waite, Gary T., 'Talking animals, preserved corpses and Venusberg: the sixteenth-century magical world view and popular conceptions of the spiritualist David Joris (*c*.1501–56)', *Social History*, 20: 2 (1995), 137–56.

Walker, Nigel, *Crime and Insanity in England: Volume One: The Historical Perspective* (Edinburgh: Edinburgh University Press, 1968).

Webster, Charles, *The Great Instauration: Science, Medicine and Reform 1626–1660* (London: Duckworth, 1975).

Welbourn, F. B., 'Man's Dominion', *Theology*, 78: 665 (1975), 561–8.

White, Lynn, Jr., 'The Historical Roots of Our Ecological Crisis', *Science*, 155: 3767 (1967), 1203–7.

Wilson, Dudley, *Signs and Portents: Monstrous Births from the Middle Ages to the Enlightenment* (London: Routledge, 1993).

Wolfe, Don M., 'Lilburne's Note on Milton', *Modern Language Notes*, 56: 5 (1941), 360–3.

Wood, Diana ed., *The Church and Childhood* (Oxford: Blackwell, 1994).
Wrightson, Keith, *English Society 1580–1680* (London: Hutchinson, 1982).
Young, Alan, *Tudor and Jacobean Tournaments* (London: George Philip, 1987).
Zagorin, P., 'The Authorship of *Mans Mortallitie*', *The Library*, Fifth Series, 5: 3 (1950), 179–83.

Index

Printed in the United States
By Bookmasters